TechnoLogics

THE SUNY SERIES IN
POSTMODERN CULTURE
Joseph Natoli, *Editor*

TechnoLogics

Ghosts, the Incalculable,
and the Suspension of Animation

Gray Kochhar-Lindgren

State University of New York Press

Published by
State University of New York Press, Albany

For information, address State University of New York Press,
90 State Street, Suite 700, Albany, NY 12207

Production by Diane Ganeles
Marketing by Susan Petrie

Library of Congress Cataloging-in-Publication Data
Kochhar-Lindgren, Gray.
    TechnoLogics : ghosts, the incalculable, and the suspension of animation / Gray
Kochhar-Lindgren.
        p. cm. — (SUNY series in postmodern culture)
Includes bibliographical references and index.
ISBN 0–7914–6303–6 (alk. paper) — ISBN 0–7914–6304–4 (pbk. : alk. paper)
    1. Technology—Social aspects. I. Title. II. Series.

    T14.5.K62 2004
    303.48'3—dc22   2004042984

10 9 8 7 6 5 4 3 2 1

for
Duncan
and
his generation

Do we stand in the very twilight of the most monstrous transformation our planet has ever undergone, the twilight of that epoch in which earth itself hangs suspended?

—Heidegger

No one is allowed on these premises if he is afraid of machines and if he still believes that literature, and perhaps even thought, ought to exorcise the machine, the two having nothing to do with each other.

—Derrida

Contents

Acknowledgments

Writing is an act of gratitude. My thanks go to Barry Alford, Anne Alton, Gary Astrachan, Bill Brevda, Mary Ann Crawford, William Doty, Hedwig Fraunhofer, Peter Haddad, Heidi Holder, Robert Hrdina, Joseph Lease, John Moore, Jerry Neeb-Crippen, Daniel Patterson, Robert Paul, Steve Scholl, David Smith, Christina Tassev, and Aparna Zambare. A special note of gratitude to Ron Primeau for initiating me into the lore of the goat, to Stacy Thompson for keeping me meandering toward the flag, and to Mark Freed for holding the nonmodern door to the Academetron open.

The Dean's Office of the College of Humanities, Social, and Behavioral Sciences at Central Michigan University, as well as the Faculty Research and Creative Endeavors Committee and the Office of International Education, provided essential support for my work. Parts of *TechnoLogics* first appeared in scholarly journals. My thanks to Arthur and Marilouise Kroker, editors of *CTheory*, to Mark Bracher, editor at the *Journal of the Psychoanalysis of Culture and Society*, and to Geraldine Shipton at *Psychoanalytic Studies* for their permission to incorporate the revised essays into the book.

It has been a pleasure to work with all those involved at the SUNY Press. The astute comments of the anonymous readers greatly improved the book. Without the invaluable assistance of Wyatt Benner, Diane Ganeles, and Susan Petrie, as well as the expert guidance of Joseph Natoli, the series editor, and James Peltz, editor-in-chief at the Press, there would be no book. My gratitude to them all.

Hats off, as well, to the kinfolk on this and the other side of the line: Prakash Chand, Chris, Anil, Marilyn, Stephanie, Rachel, Nathan, June, Gray, Tom, Laurie, Will, and Hunter.

As for Kanta, who travels between worlds, well, she knows what she does.

Introduction

The most concerned ask today: "How is the human to be preserved?" But Zarathustra is the first and only one to ask: "How is the human to be overcome?"

—Nietzsche

echnoLogics is an attempt to understand the uncanny logic that is unfolding for all of us as it enfolds all of us in its multiple orders. It concerns itself with small prefixes such as *un-*, *re-*, and *in-*, as well as with the broad movements of history that leap from the Hebrew Bible and Plato straight into the nineteenth century and beyond. Rather than directly analyze the chronological emergence of the dominance of technology, this work attempts to show that the line and all of its affiliates—progressive teleologies, certain notions of causality, the logic of identity, and, especially, the fundamental concept of time that has governed western history from the alpha to the omega—are becoming snarled, twisted, broken, and far more complicated than we had previously imagined. Superstrings are not straightforward; hyperlinks are not logical sequences. The order of the alphabet, as well as the *logos* of *chronos*, is being transformed. These complications at the end of the line, which of course does not simply vanish, have implications for all of our categories of experience, including the very possibility of subjectivity and the generation of meaning for the future of what is coming to be called the "posthuman."

In *TechnoLogics* I hope to mark this shift, and, in particular, to work along with innumerable others to rewrite the old table of values that, for millennia, has defined the fundamental opposition between the living and the dead, the animate and the inanimate, and that has governed all sorts of other oppositions. The logic of technology turns out to be not a simple instrumentality of reason—as if reason were a unified force that could be equitably *applied* to a variety of problems—

but a supplementary appearance of the uncanny, of that force which displaces certainties and changes the very nature of our being-in-the-world, within the very heart of rationality. What would happen if we were, as Martin Heidegger suggests, to seriously consider whether the "*legein*, as a laying and a letting-lie, would be something uncanny in the midst of all the canniness of human existence?" (1968, 206).

The "law" of technologics, arising from the establishment of the *logos*—and the accompanying repression of the "primitive" *mythos*—as the ground of thought, declares that the utilitarian calculability of the "clear and distinct" is always accompanied by its other, by what I am here figuring as "ghosts," which not only walk abroad across the face of the earth, but are also determining a great deal of contemporary cultural theory in sociology, philosophy, psychoanalysis, and literature.[1] Antonio Negri points to the extreme visibility of the spectral in our time, asserting that deconstruction

> produces a new theory of spectrality which corresponds with common experience: an experience of the everyday, and/or of the masses; the experience of a mobile, flexible, computerized, immaterialized and spectral labor. *A common experience of spectrality as clear as the sun.* The new spectrality is there—and we're entirely within this real illusion. . . . There's no longer an outside, neither a nostalgic one, nor a mythic one, nor any urgency for reason to disengage us from the spectrality of the real. There's neither place nor time—and this is the real.
>
> Only a radical "Unheimlich" remains in which we're immersed. (1999, 9)

Something, as they say, is afoot, and this thing is appearing wherever we look in transepochal culture, a globalized culture of calculability and its others, of spirits and spiritualities, of a "radical *Unheimlich*" which shows itself everywhere, but is without specific location. It dislocates, both temporally and spatially. Spectrality is not some esoteric knowledge given only to the adepts of deconstruction or psychoanalysis; rather, it is all around us in our most mundane activities and transactions; it is the medium in which, and as which, we exist.

With the experience of rationality itself as the uncanny come ghosts, repetition, terror, death, and the need to rethink all of our categories of interpretation. After the ascent toward the truth of

Platonism-monotheism and the descent into the infinite interiority of the self inaugurated by romanticism, we have now entered a new epoch in which we are no longer oriented by the top-bottom, inside-outside divides, but in which we slide along multiplanar plateaus of the networks of signification, which include the productivity of the nonhuman. Félix Guattari is correct when he claims that "Technological machines of information and communication operate at the heart of human subjectivity, not only within its memories and intelligence, but within its sensibility, affects, and unconscious fantasms" (1995, 4). Identities—shorthand for various relational interfaces with the environment, machinic and otherwise—are multiplied and stretched, reconfigured, and we are *suspended* between eras (although this "between" does not indicate a chronological line that begins on one side of the transepochal and picks up, as if unchanged, on the other). And this suspension has its own structure.

While it is certainly the case that linearity—as method, protocol, procedure, accounting practices, and a form of experience—continues to be absolutely essential to the development of the technical life-world in which we live, this linearity occurs within a more encompassing unfolding that disrupts the very concept of line with the irruption of the unfathomable, the incalculable. This incalculable, of course, can never show itself, except, perhaps, long after the fact, as a series of phantom traces like electron shadows, since it is not, in and of itself, representable. Nonetheless, even if we cannot capture it in the nets of traditional conceptuality, we can gesture towards both its affects and effects.

It is this gesture which I find most intriguing about certain forms of interpretive discourse. *TechnoLogics* occurs along the pathways opened by the philosophical history of the line, Plato's articulation of the (in)visible, Shelley's and Hawthorne's alchemical fictions, Marx's linkage of capital and the machine, Melville's examination of not-logic, Jünger's manufactured garden full of the buzz of bees, Freud's analysis of the soulful apparatus, Heidegger's questioning of the world as picture and as standing-reserve, Lacan's surveillance of the machinery of planetary surveillance, and Derrida's deconstructive yes of textuality. These proper names are, of course, nodes along much more extensive expressways of citation, but they are the ones to which I am most deeply indebted. Rather than write another direct commentary on these thinkers, however, I prefer to *inflect* and *accent* their work in certain directions that might help us understand *technē* and the uncanny, the calculable and its others, the forms of temporality and (post)human identity that now seem to be emerging. And since the uncanny, the unconscious, the un-thought, and *différance* can never present themselves—there is no proper self to present—all we can do is to undertake *readings* of specific texts and letters that can never be taken literally.

Thinking, as reading/writing, is always figural, and thus one name for the task set for all of us is *literature*, the very discourse of the figural itself. Derrida, along with many others from our own as well as earlier historical periods, has formally breached the divide between the philosophical and the literary, that form of writing that is always suspended between the true and the false, that can never make the same sorts of claims as history, science, and meta-physics, but that—like an uninvited guest—appears to trouble the dreams of those other disciplines. Disciplines, as we all know, con-tinue to exert real force and are the ongoing necessary fictions of institutional life, but, as excluding forms of thought and practice, their day—as they have been constituted since the rise of the Ger-man model of the research university—is coming to a close. The *epistēmē* of the late twenty-first century will look quite differently than that of the late twentieth and we are now, already, beginning to make that transition.

In *TechnoLogics*, then, I will read texts considered literary, philo-sophical, and psychoanalytic, with the technosciences and techno-capitalism always in the vicinity and sometimes, especially in the sections on Marx, Lacan, and Jünger, brought into the foreground. Always, I will try to situate myself in the gray zone between the living and the dead, attendant to the approach of machines and ghosts—how, if at all, do they differ from one another?—and hoping to learn something from speaking to phantoms and listening to voice mail.

After "Call Forwarding"—which is a brief reflection on finitude, the wish for immortality, and technology—I will, in "On-(the)-Line," play with the line as it has been cast into language as a figure for all sorts of things, most pertinently, time. In this chapter, I will examine how thought, as the emergence of metaphysics, moving from the so-called conceptual to the so-called material, laid down the conditions for the eventual literalization of the power grid of the technological world in which the human is completely saturated. Passing the line through Plato, Aristotle, and Kant, I will then analyze Heidegger's *Über die Linie*, which began as a birthday greeting to Ernst Jünger and raises the question of the line and nihilism. Derrida's line is then cast, and I will begin what cannot be finished, but that can be echoed and reiterated: an unwinding of the time line as we all come on-line.

In "The Platonic Teleport," I examine the establishment of the name of "Socrates," which is peeled away from the empirical man of that same name, as the transmitter of "philosophy" to the future, and show how, already, rationality is imagined as transcending the division between life and death. Through a reading of the *Cratylus*, and with extensive help from the Derridean critique of *mimēsis* and iterability, I suggest that cloning, too, is nascently present in the machinery of the Platonic dialectic as it attempts to abstract and mathematicize itself on the ascent toward the infinite day of life and light devoid of death and darkness.

On its 2002 phone-cards, France Telecom beautifully articulates this wish to obliterate the cave: "Le téléphone ignore la nuit, / il est le / jour infini." As if glossing this télécarte-poem, which is the poem of the dream of absolute daylight, Jean Baudrillard reminds us of the beneficial necessity of the gap between a star's light and its delayed perception by us, since the "simultaneous perception of the light of all the stars would be equivalent to an absolute daylight, and this would be unbearable for us" (2000, 72). The fantasy of an infinite day or of a finality of simultaneity—information without static that is available the instant we desire it—entails a complete erasure of the vacillation between day and night, between the moment of desire and its (non)fulfillment, and of the rhythmicity upon which both time and writing depend.

This chapter concludes with a discussion of the entry into spectrality that announces itself, if such can be said, at the end of the line when philosophy becomes what it has always latently been: cybernetics. Plato's writing, then, acts both as a sorting-machine of values, drawing up the schema of the divided line with its separation of *psyche* and *soma* that will haunt idealism, including the idealism of the technological sublime that will want to cast off the body for the sake of the durability of pattern and information, as well as a facsimile-machine that will transmit forward into history its own graphing of being.

Once this thought-grid of technical value and process is laid out, I will pick up the Platonic fax at three different nineteenth-century destinations: in the alchemical laboratories of Nathaniel Hawthorne and Mary Shelley, whose protagonists are seeking immortality through the wonders of chemistry; in the *Werkstelle* of Karl Marx, where human beings are forged into the social-technical intricacies of the machine; and, finally, inside the desk drawer, among the ginger nuts of the law, of one Bartleby of Wall Street. As the Platonic message of rationality, duplication, and naming has been distributed, rerouted through countless servers—translators, libraries, economies, religions, sciences—it has, of course, changed its message. All reception, after all, transforms any transmission, and we will attempt to decode the way the Platonic text arrives in these other mailboxes.

It's not that the fax didn't arrive at destinations before the nineteenth century. It, of course, did. And it's not that technology, science, and philosophy do not have extraordinarily rich histories from Plato until the nineteenth century. They, of course, do, and others have told those tales masterfully. But it is in the nineteenth century that capital, technics, and spirits meet in a new and powerful configuration that programs the social forces that launch the twentieth-century unfolding of the technosciences, which will be joined at the hip with technocapitalism.

In his analysis of Marx's magic table in *Capital*, Derrida explicates what is already present throughout the nineteenth century: the fact that "the capital contradiction does not have to do simply with the incredible conjunction of the sensuous and the supersensible in the same thing; it is the contradiction of automatic autonomy, mechanical freedom, technical life. Like every thing, from the moment it comes onto the stage of a market, the table resembles a prosthesis of itself" (1994b, 153). These "contradictions," the dialectically irresolvable paradoxes of an emergent logic of the suspended, always generate, and are generated by, their own "prostheses"—their own supplements, mirrors, and ghosts. The "real," therefore, can never be purified of its own haunting spirits, and, if the latter are in fact one day obliterated, the "real" will vanish as well.

The instantiation and institutionalization of the machine that is everywhere present in the nineteenth century may *wish* to exorcise, for the sake of rational efficiency and a rational ego-ideal of subjectivity, all the ghosts of the past, of tradition, religion, the primitive, the irrational, and the mythic; but, instead, the ghosts—as the uncanny always does—return within the very structural heart of the machinic. It is not that there is the ghost of the spirit *within* (and therefore separate from) the exterior shell of the machine, but, rather, that the machinic *itself*, the *Gestell* as a whole, is ghostly, haunted by all that it attempts so violently to repress. Avery Gordon is correct when she asserts that

> Haunting is a constituent element of modern social life. It is
> neither pre-modern superstition nor individual psychosis; it

is a generalizable social phenomenon of great import. . . . The way of the ghost is haunting, and haunting is a very particular way of knowing what has happened or is happening. Being haunted draws us affectively, sometimes against our will and always a bit magically, into the structure of feeling of a reality we come to experience not as cold knowledge, but as a transformative recognition. (1997, 7–8)

I will begin part 2, which traces the ideology, the interpretive construct, of technics and capital in the nineteenth century, with a brief presentation in "The Elixir of Life" of two short stories: "Dr. Heidegger's Experiment" (1837) by Nathaniel Hawthorne (how could I resist?) and "The Mortal Immortal" (1833) by Mary Shelley (with Dr. Frankenstein always lurking in the back room). In both cases, alchemy is employed in hopes of creating immortality, with its attendant problems, as science reaches beyond the limit of death. Philosophy and religion, as we will see with the desire of Socrates, has always made this leap, but now it is the physical sciences that begin the attempt to literalize the mythic-conceptual-metaphorical order of its ancestors.

In "The Immortality Machine of Capitalism," I will examine Marx's stunningly prescient analysis of the social-machinic in the *Grundrisse* (1857–58), where the old chemistry of the *lapis philsophorum* becomes the "alchemical cauldron of capitalism" that transforms everything that enters its alembic into a commodity that has exchange value. Laboring to exorcise the Hegelian *Geist*, Marx analyzes the vampiric nature of capitalism as its machinery, apparently dead, devours the bodies of its workers, apparently alive. Already, by midcentury, the world was coming on-line. It was not yet coupled through the assembly line of Frederick Taylor and Henry Ford, much less by electronic interfaces, but it was a site, means, and product of industrial labor that prepared the social ground for the advent of the scientific-capitalistic globalism of the next century.

In "Bartleby the Incalculable," I will construct an interpretation of the capitalist nexus of Wall Street with its attendant ghost, Bartleby, who knows only, like Socrates, how to say the "not," a sign of resistance that may turn out to be the last bulwark against the disappearance of the human as the human. This very little story, this "almost nothing" as the narrator will call it, is for me the nuclear poem of the nineteenth century; its extraordinary compression expresses all of the major thematics of the period—indeed, of metaphysics itself. In my analysis I will explicate some of these themes, especially

as they open up the "day and night line" of capitalism, the presence
of the wight at the heart of the financial and legal institutions of the
country, and how the scrivener raises the question of the nothing as
the incalculable. What else can result but a death at the base of the
monumental wall of the pyramid? Perhaps another reading, and, in
any case, a letter—whether living or dead is not easy to ascertain—
sent into the next century.

In the final part of the book I will examine the twentieth-
century consequences of the coming on-line of human existence
that has been in the making since the voice-overs and ventrilo-
quisms of the biblical and the Platonic corpus. In "The Drone of
Technocapitalism," a reading of Jünger's 1957 novel *The Glass
Bees*, I will show how capitalism, cloning, the media, and war con-
join to create a simulated world for the sake of production and
profit. Best known for his dialogues with Heidegger, his memoirs
of World War I, and for his analysis of "total mobilization," Jünger
has written a novel whose protagonist, Zapparoni, may or may not
be "real," but nonetheless controls a media-financial empire that
has radically deepened the incursion of the powers of biotechnical
fabrication into the natural world. The world of Zapparoni is the
anti-Bartlbeian world, for its titanic wager is that nothing, not
even the nothing itself, can resist being absorbed by the manufactured world of images and artificially reproduced nature. The
"prefer not" is absorbed, and therefore nullified, by the powers of
an absolutely controlled social world as Captain Richard, the narrator, moves from having been made redundant—as both cavalry
officer and tank inspector—to becoming a temp for the Zapparoni
Works, Unlimited.

In "The Psychotelemetry of Surveillance," I continue to probe
the world of social control, surveillance, and the "soulful machine"
of the psychic apparatus designed by the history of "Oedipus" as it
is translated into (post)modernity by Sigmund Freud, J. J. Goux,
and Jacques Lacan. After establishing the multiple relay points of
that enigma called "Oedipus"—which bears directly on the emergence of the one-who-knows through the application of a rational
methodology based on the threat of violence and ending with
death—I show how Freud creates the figure of Oedipus as a unified subject of desire that spans genres and historical periods, and
how his story as such, is one of the founding narratives of technologics. In *Oedipus/Philosopher*, Goux argues that the Theban exile-
king is the first modern, as he leaves the mythic orientation of
symbolic initiation behind for the sake of knowledge that he,

himself, produces as a necessary procedure for self-knowledge linked to science and politics.

The psychoanalytic project, dependent as it is on the unconscious and its nonrepresentability as a thing-in-itself—and thus always dependent on a temporal detour of signification—disrupts the wish of immortality and the overcoming of nature that drives the technologic. Lacan, along these lines, will help us think about the field of culture as a whole, and, more particularly, about how a certain surveillance system is established, on a planetary basis, within that field. I argue that an oedipalized superego, always obvserving itself and its others, has set itself in place as overlapping systems of surveillance, including eavesdropping on all electronic transmissions, the installation of video cameras at every corner (as well as within what used to be called the "private sphere"), and the cyborgification of the body itself.

In "Temps," a play on time, work, and on the turning of the millennial weather, I begin to draw the previous chapters together, especially in terms of how they bear on the question of the temporality of technicity. Conjoining a Heideggerean exegesis of the *ecstases* of temporality with a psychoanalytic laying-out of time as *Nachträglichkeit*, a time of deferral and recursive loops, I will bear down on the logic of suspension, which suspends animation—as the living soul—along with everything else. I will attend to the blink, the flickering, the stutter, and the dark spot at the heart of the signifying process, and, once more, return to the presence of the uncanny that courses along the networks of all technologics. I loop back to the notion of iterability and *mimēsis*, connecting them once again to the entanglement of the time line with all of its consequences for the emergence of the posthuman. We are, I suggest, reconfiguring the configuration itself.

In the conclusion, "Heeding the Phantomenological," I attempt to wrap things up by opening things up, by unwrapping a present. What is to come of us? What is it like, and what will it be like, to be the posthuman? I lay out various scenarios and focus, one last time, on the blink and the echo, the delay, whatever it is that both founds and resists signification. I suggest that the dream of an infinite day is dangerous in the extreme, and that, as an alternative, we should attend to the mysteries of shadows, nuance, and night; that we might do well to listen to the phantoms that beset us, perhaps even inviting the unhomed into the heart of our discourse; and that *reading*, understood phantomenologically, is an art yet to be learned. As we move more deeply into the posthuman that we are, we will have to elaborate a technopoetics, a mythotechnics that will take account of, without ever completely accounting for, both the rational and its multiple others. We will have to learn, all over again, how to dream, how to think, and how to write. As if that were possible.

TechnoLogics ends, then, with a countdown that is set, as if automatically, to head toward the click of zero so that we can, perhaps once again and perhaps for the first time, listen to the machine coming to life—

I

Laying Down the Power Grid

The vertigo of a world without flaws.

—Baudrillard

1

Call Forwarding

The movement of progress, or if one prefers, the movement of process, will endeavor to reconcile values and machines. Values are immanent in machines.

—Guattari

Welcome to your voicemail service.
Please enter your password and press #.
You have x new messages and $x+1$ saved messages.
(Who knows their origins?)
Press 1 to hear your messages,
2 to change your greeting.
For wake-up service, press 3.
To repeat this menu, press *.
(What voices speak these messages?)

Cloning, the immortality of the reproduction of the same, has always been the openly secret dream of *homo phantasticus*, but such dreaming, in the dominant history of the west, has been subsumed by the logic of the drive for the rational explication and use of the world that we have come to call metaphysics. This subsumption has not destroyed the dreaming, but, rather, has installed the methods of cultural work necessary for its literalization. The dream has, through a vast array of pragmatically interpretive practices, been technologized. Eclipsed as a form of phantasmic text, even as the effects of its desire have grown gargantuan, the most an-

13

cient of dreams has been forwarded through the technosciences and technocapitalism into the smallest niches of everyday life and into all the tissues of our bodies. The posthuman, although we will have to articulate different versions of this phenomenon, is the culmination of a certain enacted interpretation of a species-dream. We need to think about how to forward that dream onward and, perhaps, how to entangle it, send it on incessant detours.

Re-production is the center of this pragmatic interpretation that we call, usually too simply, the "history of the west." The wish that drives the dream is to re-place the given with the made, the so-called natural with the artifactual, and, thereby, to control the *re*-of production. This, along one of its lines of development, entails the destruction of gendered sexual reproduction and an incorporation of the body of the mother of nature into a technological process. This process appears in a number of recent novels organized for the most part around the genre of cyberpunk, as well as films such as *Bladerunner*, *Alien*, *Gattica*, *The Matrix*, *Minority Reports*, and *AI*, but I will use the names of "Socrates," "Oedipus," "Bartleby," and "Zapparoni"—among others—to try to deepen our understanding of the dynamics of the emergence of the posthuman. Rather than destroy the (in)animate as a category of existence, the posthuman continues, instead, to put it into *suspension*.

Cloning is the fantasy of the reproduction of the same through technical means of production, and, as Derrida has noted, "The *eidos* is that which can always be repeated as the same. The ideality and invisibility of the *eidos* are its power-to-be-repeated" (1981, 123). It is such repetition that provides the condition for the possibility for the re-cognition of any identity (whether of logic, language, tradition, or person). In the contemporary world of global capitalism and of the sciences that indissolubly meld *bios* and *technē*, such repetition is also being broached not only as the invisibility of the ideal, as that form of rational thinking itself called philosophy, but also in the order of materiality, where instrumentation has enabled the invisible to become visible and to be worked upon for human ends. The displaced synonyms for the *eidos* are genes and computer code.

The interiority of the invisible, from this perspective, has actually only been in hiding and now comes to the light of rational-empirical inquiry as the exteriority of the visible and therefore the usable. But just as nature, following Heidegger's argument in "The Question concerning Technology," becomes a standing-reserve for our apparent enhancement, so we, too, become part of that standing-reserve, that inventory of commodities, that machine-assemblage of global culture.

This does *not* mean that we, as free ideal beings of spirit, are somehow constrained, captured against our will by the machine-assemblages in one version or another of the iron cage of materialism. It means that the very form of our being, in the simplest form of daily life, is to be technobios, to be cyborg.

This, in itself, is not a new development. The spirit has always been machined—there have always been techniques of the sacred—and the machine has always had a consciousness component to it. As Derrida so forcefully reminds us, there is no language, and thus no humanity, without the machine of repetition and transmission—of voicemail and faxes—embedded in the productions of meaning. The *psyche-soma* has always been an apparatus, a magic-writing pad, an automatic archive. Now, however, in the transepochal, the lines have been crossed in a different manner, with higher stakes.

In Baudrillard's terms, we are undertaking a project

> to reconstruct a homogenous and uniformly consistent universe—an artificial continuum this time—that unfolds within a technological and mechanical medium, extending over our vast information network, where we are in the process of building a perfect clone, an identical copy of our world, a virtual artifact that opens up the prospect of endless reproduction. (2000, 8)

One of the consequences of this move in which difference is erased—and to which we shall return in both "Temps" and in "Heeding the Phantomenological"—is that the world has become "so accelerated that processes are no longer inscribed in a linear temporality, in a linear unfolding of history. Nothing moves any longer from cause to effect; everything is transversalized by inversions of meaning, by perverse events, by ironic reversals" (2000, 78). The crossing of lines marked by an exponential increase in the acceleration of events reconfigures the time line, history, causality, the thing, the commodity, and the meaning of the human. This, in turn, reconfigures the logic of all dichotomies, releasing—if that's the word—the uncanny.

The enfolding of all binarities is appearing, and, for the time being, we exist in the creases of this appearing which constitute us by dividing us in other ways than body and soul, life and death, nature and culture. Both the similarities and the differences at work in this "and" require explication. We must therefore continue to think, alongside others but with different tonalities, the logic of suspension, a logic that does not destroy old antinomies but keeps them in tension with one another. One, and other. Othering the one. 1, 2; ½; 1×2; 1=2. "All would be simple if the *physis* and each one of its others were one or two. As we have suspected for a long time, it is nothing of the sort, yet we are forever forgetting this. There is always more than one—and more or less than two" (Derrida 1996, 2). Identities are multiplying, and all multiplication, the many foldings of cultural textuality, creases the concept of identity.

. The motive of the reproductive wish is the wish to escape death, whatever it takes. The speaking god of Genesis makes human beings in its own image, and then wants that image to multiply on the face of the earth while remaining faithful to the most primordial image of the *imago Dei*. Plato aligns truth, virtue, and immortality with the proper intellectual imitation of the forms. And, yet, there is always a snag. Yahweh, taking command over the other name of the creative voice in Genesis 1, jealously refuses the tree of immortality; banishes his image to exile, suffering, patriarchy, and death; and confuses the languages of the earth. For Plato, just across the Mediterranean, there is always the danger of bad copies, shadowy images that taint the clarity of the original. People might, he worries, prefer the pleasures of the cave to the asceticism of philosophy, and Sophists, he fears, lurk everywhere, with love letters or political polemics nestled in the dark depths of their robes.

Postmillennial technocapitalism, which is unthinkable without science, wants to smooth out those minor snags, to pass back across the boundary line traced by the flaming sword and teach everyone to speak the same language. It works assiduously, night and day, to make sure that copies, free of noise, can travel anywhere at any time. It produces the Book of Life as the human genomic code and wants all codes, the deeply cryptographic nature of nature, to become unscrambled and readable for the sake of utility. Nature might love to hide, but now we often assume that we have opened the master locks of its treasury. As Gilles Deleuze states, speaking of the "control societies" in which we live,

> codes are passwords [and] the digital language of control is made up of codes indicating whether access to some information should be allowed or denied. We're no longer dealing with a duality of mass and individual. Individuals become *"dividuals,"* and masses become samples, data, markets, or *"banks."* (1995, 180)

Those with the secret entry code can gain access to the dataworld, which is tantamount to the power of information, and the technocharged installation of Being wants to be the end of all childish games like hide-and-seek. This, after all, is serious business, and video games are intimately related to war games.[1]

For us, living at the pivot of the transepochal, the most primordial of dreams have been concretized as cloning, genetic engineering, artificial intelligence, and cybernetics in its many guises. The logic of the same roars ahead, unabated and turbocharged. Stand behind the yellow line, please. Mind the gap. Keep back, but stay put. This train is computerized; it keeps its own schedule and will arrive any minute now. Everything, after all, is being tracked.

We have a moment, but only a moment, to talk, to learn to listen. To speak out.

There has always been a reproduction in which the doubled, the living dead or the dead-in-life, returns to haunt the present with its uncanny presence. As Baudrillard so succinctly puts it:

> Of all the prostheses that mark the history of the body, the double is doubtless the oldest. But the double is precisely not a prosthesis: it is an imaginary figure, which, just like the soul, the shadow, the mirror image, haunts the subject like his other, which makes it so that the subject is simultaneously itself and never resembles itself again, which haunts the subject like a subtle and always averted death. This is not always the case, however: when the double materializes, when it becomes visible, it signifies imminent death. (1994, 95)

We are faced with an imminent death—though this may also be a moment of promise—and the double from the other side, which in our age becomes connected to the prosthetic, used to be called a ghost or a *Doppelgänger*; but now, in an epoch of digitized engineering, it is called robot, cyborg, clone. We have crossed the line that once attempted to secure the difference between the animate and the inanimate, the so-called living and the so-called dead, and once that line is crossed, twisted, and folded, there is no going back. As Katherine Hayles describes the situation:

> The great dream and promise of information is that it can be free from the material constraints that govern the mortal world. Marvin Minsky precisely expressed this dream

> when . . . he suggested it will soon be possible to extract
> human memories from the brain and import them, intact
> and unchanged, to computer disks. The clear implication is
> that if we can become the information we have con-
> structed, we can achieve effective immortality. (1999, 13)

It is essential that we not only analyze the categories of "dream" and
"promise," and their profound filiation with the hope for immortality, but that, in addition, we *produce* a variety of models for the new
languages of dream and promise with which we are confronted
through the logic of *technē*. We who inhabit this millennial turning
live neither on one side of the line nor on the other: we are suspended
between the two, a vacillation between forms of being, neither of
which we belong to in our essence. This allows us, for the moment,
to look both ways.

   We are the crossing, the uncanny juncture where the track of
the most primitive dreams meets the most modern of the powers of
the rational, and this is the location where *TechnoLogics* sets itself
up as it traces one trajectory, and only one, of the history of these
crossings. It is the story of rationality and the vicissitudes of philosophy and its incestuous others, literature and psychoanalysis; of
the emergence of machines from within, and as, the *Zeitgeist* of
time's soul; and of the new entities, ourselves included, who move
now among us, alien with promise and fear. It is a story of nothing,
of duplication, of ghosts.

   It is the story of a call and a response to that call, for there is
always someone, or something, calling us to think.[2] That may be
number, a wight, or something altogether different. And, for the
task before us, a kind of dreaming-thinking might be required, or,
at the very least, a recognition of the hyperlinks between what
Freud called, with his usual simplicity, the primary and the secondary processes and all their analogues. Living along these links
will require a new kind of *poesis* that re-visions the relationship
between the rational, the nonrational, the affective, the mytholog-
ical, and the critical. This is probably an impossible task, but
must, in any case, be taken up from a host of different sites, in a
host of different idioms.

   As we shall see, Plato heard the summons to thought as a call
from the digits, as both fingers and numbers. But now, at the end of
metaphysics and its transmutations into its nano- and macroforms,
the call sounds differently. Summing up an entire history, Heidegger
writes that

Just because at one time the calling into thought took place in terms of the *logos*, logistics today is developing into the global system by which all ideas are organized. . . . Western logic finally becomes logistics, whose irresistible development has meanwhile brought forth the electronic brain, whereby man's nature and essence is adapted and fitted into the barely noticed Being of beings that appears in the nature of technology. (1968, 163, 238)

As organic beings, we are being (retro)fitted by the essence of technology even as we fit all of nature to fit in with our demands for knowledge and resources. We hear a kind of double call. On the one hand, we continue to be seduced by the call of digitization and its analogues. We must continue to forward the Platonic call as a pattern, a program, that will fit materiality to its code. On the other hand, there is a persistent effort underway to listen for another call and to make a gesture toward an "other metaphysics" that might appear unexpectedly, outside the parameters of the Platonic program. But "on the one hand and on the other hand" assumes only two hands signing the conceptual move from logistics to dialectics, and we have entered the period of the multiplication not only of hands, but of all other body parts as well.[3]

It is the cyborg, which is always already a phantasmic entity, that calls to us, through us, and for us. How will we respond? Is it possible to be responsible, and, if so, in what sense?

We still have time, a little morsel of time, before we have to go. It gives us, as the law gives the one who waits for death in a standing reverie, food for thought.

Push start. Wait a moment.

Augenblick.

Just the blink (and the blink will always be repeated) of an eye. The darkness descends; light (re)appears.

Nachträglichkeit and *Ereignis*.

Why should we wait? What should we wait for? What's the delay?

The machine will begin to purr.

It already has, long ago, almost as if running on its own—

2

On-(the)-Line

There is a low humming in the background.

Everything is now on-line. Powered up.

We must, whether we want to or not, put ourselves on the line as well.

What, though, is a line? What is the course of the line between the living and the dead, the animate and the inanimate? What does it mean that now, driven by the program of technology, we are attempting crossings back and forth across that line that has long been prepared for, but never before, not in the same way, driven along?

What is a line? A mathematical stratagem. The shortest distance, we have been told since we were children, between two points. Or, to put it a bit differently, a line is the *Aufhebung* of the point:

> As the first determination and first negation of space, the point spatializes or *spaces* itself. It negates itself by itself in its relation to itself, that is, to another point. The negation of negation, the spatial negation of the point is the LINE. The point negates and retains itself, extends and sustains itself, lifts itself (by *Aufhebung*) into the line, which constitutes the *truth* of the point. (Derrida 1982b 42)

The point is that the spatializing point produces the line of time, one point after another as one "now" gives way to the next, but this, Heidegger will argue, is but the "vulgar conception" of time, a contention that will not remain unchallenged.

A line is a boundary that marks off divisions of the land, like a cadastre, a survey of the topography like the one that will occur in "Bartleby." The surveyed is the surveilled, and everything, all too soon, will be measured and positioned beneath the grid of observation outlined by the eyes in the sky and by the network of closed-circuit cameras that are becoming an invisible part of our urbanscapes.[1] It's as if the lawyer's oversight of his guest, the one who says no, has been projected into an entire system of watchfulness, but one that can see only objects within a certain set of parameters and will, out of a structural necessity, miss its own blind spots. Continuous oversight, we might say, is blind from the beginning. And, as I will suggest, Oedipus, who mans the line of western metaphysics, is one name for the operator of this system.

Lines multiply. A lifeline runs across the palms of our hands and we've all pondered that short, curved indentation of the flesh. The line is always one of life and death, their codetermination. Lines are always connectors—drop me a line when you get a chance; I think of you often—but the lines can, always, go dead. The figure of the line combines technologies of communication, the alphabetic and the electronic. There is always, of course, a small hookup fee, a small *unbehaglich* tax—the merest sliver of the pleasures of the flesh—and it gives power, a buzz that lights up the night.

The line is a command, but is a line up really needed? Are these really criminals we're looking at? In a laboratory of Plato or Cornelius Agrippa, in the offices of lower Manhattan, the reading room of the British Library, in an old Cistercian monastery, or in the watchtowers of respectable society? Thugs numbered and profiled in silhouette? Are they part of the mob? How can we identify the right one, when we're not sure what, or even if, crime has occurred? And what lineup are we standing in? The disaster has occurred; we all know that

> The devastation of the earth can easily go hand in hand with a guaranteed supreme standard of living for man, and just as easily with the organized establishment of a uniform state of happiness for all men. Devastation can be the same as both, and can haunt us everywhere in the most unearthly way—by keeping itself hidden. Devastation does not just mean a slow sinking into the sand. Devastation is the high-velocity expulsion of Mnemosyne. (Heidegger 1968, 30)

But is this a *crime*? Mnemosyne has been out of fashion for quite some time. Can someone, anyone, be held responsible? What are the roles of agency, ethics, and memory in the transepochal shift we are undergoing? Especially if "agency" is reexperienced along the lines Andrew Pickering suggests when, discussing the actor-network theorists of the sociology of science, he argues that "there exist important *parallels* between human and material agency, concerning both their repetitive quality and their temporal emergence; and that a constitutive *intertwining* exists between the material and human agency" (1995, 15). But although there are in this account emerging symmetries between human and nonhuman material agency, it is the "extended temporal sweep of human agency" (19) in which the symmetry breaks down. Time, tied into an interpretation of intentionality and the formation of the future, still provides, at least in this instance, the line between the human and its others. But as "we" take up different places in the network of the (non)human, it seems, too, that the sense of temporality would undergo a marked shift. What will this new "keeping time" look like?

How many voices are *on* this line? It seems to run off in all directions, toward the vectors of all the forking paths that lead to an "infinite series of times, a growing, dizzying net of divergent, convergent, and parallel times" (Borges 1964, 28). Why is it so often garbled, and how many languages are flowing through the circuitry, the servers? I hear ghostly murmurings. Philosophers are whispering to each other in the dark, and they are all obsessed with lines. In the *Physics*, Aristotle mutters that

> The now is a link of time, for it links together past and future time, and is a limit of time, since it is a beginning of one and an end of another. But this is not manifest, as it is in the case of the point at rest. It divides potentially, and *qua* such, the now is always different, but *qua* binding together it is always the same, just as in the case of mathematical lines. (*Phys.* 4.222a.49)[2]

The now, the present, is a link and a limit; it is but a limit only potentially, always in differential motion, and a link between the has-been and the not-yet that is always the same. The now is differentiating sameness. "Just as"—and what else can be done than to *ana-logize* (at least until digitization comes along)—the point is to the line. Time points to the line.

Jumping all too quickly along the switchboard of the history of philosophy, Kant, too, has a little something to say about the line of time. "Time is nothing but the form of inner sense," he writes, that is, of the intuition of ourselves and of our inner state. It cannot be a determination of outer appearances; it has to do neither with shape nor position, but with the relation of representations in our inner state. And just because this inner intuition yields no shape, we endeavor to make up for this want by analogies. We represent the time-sequence by a line progressing to infinity, in which the manifold constitutes a series of one dimension only; and we reason from the properties of this line to all the properties of time, with this one exception, that while the parts of the line are simultaneous the parts of time are always successive. (1965, 77)

Time is a "nothing but" without shape, but out of this lack we give it the shape of a line. Wor(l)ds bloom within the opening space of the *ana-logos*, but, for this limited now, time is depicted as a line. Then, with a slight sleight-of-hand, we *reason* from the line—time's poetic shape—back to time, with "one exception": temporal succession (of past, present, future) replaces "graphic simultaneity" (a term Samuel Weber develops).

The *logos* of the time line emerges from the *ana-logos*. In this particular trope, however, there is, in addition to the advantage of an image of infinity, also a loss: the manifold, all that might appear to us at any time in any place, is reduced to "one dimension." The multidimensionality of time, the "bouquet of time" (Green 2000, 161), has, as it were, been violently compressed by the history of metaphysics, by one reading of technologics, into a single thin line, a linked and limited row of points stretching away, in both directions, into the distance.

Next in line, Hegel (with his usual good humor) argues in the *Encyclopedia* that

Negativity, as point, relates itself to space, in which it develops its determination as line and plane; but in the sphere of self-externality, negativity is equally *for itself* and so are its determinations; but, at the same time, these are posited in the sphere of self-externality, and negativity, in so doing, appears as indifferent to the inert side-by-sideness of space. Negativity, thus posited for itself, is Time. (1970, 34)

Time, as it were, is space put into motion (but Hegel is very clear that the two cannot be thought apart from each other). He extends his presentation through a discussion of the "*Now* which, as singularity, is *exclusive* of the other moments, and at the same time completely *continuous* in them, and is only this vanishing of its being into nothing and of nothing into its being" (37). Any reflection on time will, eventually, encounter the passing of being and nothingness into one another, since coming-into-appearance and vanishing seem to belong to the essence of temporalizing.

From this point—and this is particularly pertinent to the calculability that governs the transepochal—Hegel turns to mathematics (again, a common move when time is the subject). In a statement that I take to be quite close to Heidegger's notorious "science does not think," Hegel remarks that "it would be a superfluous and thankless task to try to express *thoughts* in such a refractory and inadequate medium as spatial figures and numbers, and to do violence to these for this purpose" (38). He is attempting to ensure that mathematics, the science of number, and philosophy, the science of language, retain their distinct spheres of influence, at least at this "elementary" level of the dialectical process.

Developing his argument about the dimensionality of past-present-future (and since he equates the past with Hades, the shades would soon appear), Hegel continues his critique of the calculable, asserting that

> The name of mathematics could also be used for the philosophical treatment of space and time. But if it were desired to treat the forms of space and the unit philosophically, they would lose their peculiar significance and pattern; a philosophy of them would become a matter of logic, or would even assume the character of another concrete philosophical science, according as a more concrete significance was imparted to the notions. Mathematics deals with these objects only *qua quantitative*, and among them it does not—as we noted—include time itself but only the unit variously combined and linked. (40)

For Hegel, then, the mathematical unit is, quite mysteriously, *linked* to time, but is not, in and of itself, time. Against Hegel's attempt to maintain distinctions between mathematics and the language of philosophy, Michael Heim has written that his contemporary, George Boole (1815–64),

> inverted the traditional relationship between direct and symbolic languages. He conceived of language as a system

of symbols and believed that his symbols could absorb all logically correct language. By inverting statement and symbol, Boole's mathematical logic could swallow traditional logic and capture direct statements in a web of symbolic patterns. Logical argument became a branch of calculation. (1993, 17)[3]

Heim links this development of modern logic to the productive processes through which, eventually, language is transformed into a "network of symbols that could be applied to electronic switching circuits as well as to arguments in natural language . . . and what impact this modern logic has had on everyday language and thought is still an unanswered question . . ." (38). In his reading, this development has as its basic fantasy "No temporal unfolding, no linear steps, no delays . . . the temporal simultaneity, the all-at-once-ness of God's knowledge serves as a model for human knowledge in the modern world. . . ." (38). The premodern is linked to the postmodern through the mediation of the development of modern mathematical logics. Technologics shows itself to be, under the scrutiny of an intellectual archaeology, a compression of histories, a form of social condensation that operates much like a dream. Modernity, which transmutes the ancient as it transmits it toward the transepochal, is an *agon* between the language of calculability, which "swallows" and "captures" without delay, and its other. What will language become if its contours are shaped exclusively by the *ratio* of technologics?

Coming ever closer to our own enigmatic present, Martin Heidegger, in 1955, sent a little birthday letter to Ernst Jünger with the title *Über die Linie* (Concerning the Line).[4] When he published it in freestanding form, he retitled it *The Question of Being*, with Being understood to be "*the* question of metaphysics" (1958, 31). The line, always already, is linked with Being, nothingness, temporality, and the entire tradition of metaphysics. His letter "would like to think ahead to this place of the line and in that way explain the line. Your estimate under the name of *trans lineam* and my discussion under the name of *de linea* belong together" (36). Across, on, over, beyond, along: they all belong together on the line.

Jünger had suggested that the line indicates the "zone of nihilism" that rules modernity, and Heidegger, in response, asks:

What is the situation with regard to the prospect of a crossing of the line? Are the human component realities already in transit *trans lineam* or are they only entering the wide field in front of the line? But perhaps we are being held spellbound [*bannt uns*] by an unavoidable optical illusion. Perhaps the zero-line is suddenly emerging before us in the form of a planetary catastrophe. Who will then still cross it? And what do catastrophes do? (36)

What kind of magic might break the spell of the power of the eye and its illusions, its gazes and desires? Further along, he suggests that it is metaphysics itself that acts as a barrier to crossing the line of nihilism, and that any attempt at crossing will demand a "transformed relationship to the essence of language" (71). He notes that we are all too tempted to evaluate language according to the "tempo of calculating and planning which directly justifies its technical discoveries to everyone through economic success" (73). There is a tempo to language that, while we are trying to measure its metrics as the speech of technologics, nonetheless also calls as the immeasurable, the incalculable. This knot of temporality, language, the question of the zero, and the catastrophe will recur.

It is not, however, as if the line is something "out there," apart from the uncanny essence of *Dasein*'s existence, for "Man does not only stand *in* the critical zone of line. He himself, but not he for himself and particularly not through himself alone, *is* this zone and thus the line" (83). We are all on the line as the line. The line is the (not)place of technologics, (im)passable without an encounter with the incalculability of the unhomed ghost, a transformation of language, the development of new ears, and a different comprehension of temporality. This is the most arduous of tasks, and the fact that any attempt "to achieve that transformation presumably will still remain unsuccessful for a long time is not an adequate reason for giving up the attempt. The temptation is especially close at hand today to evaluate the thoughtfulness of thinking according to the tempo of calculating and planning which directly justifies its technical discoveries to everyone through economic successes" (73). Technoscience and technocapitalism come along with each other, and both are governed by a certain tempo, a certain form of temporality. It's a rush.

Discussing the time line of metaphysics from Arisotle to Heidegger, Hardt and Negri claim that

Time has continuously been located in this transcendent dwelling place [of measure]. In modernity, reality was not

conceivable except as measure, and measure in turn was not
conceivable except as a (real or formal) a priori that corralled
being within a transcendent order. Only in postmodernity has
there been a real break with this tradition . . . time is no
longer determined by any transcendent measure, any a priori:
time pertains directly to existence. (2001, 401)

While the issues seem to me more complicated than this, a concern
that I will address more thoroughly in "Temps," these commentators are correct to connect time, metaphysics, measure, the transcendent, the tradition, and (a very long) modernity. And, while I
am distrustful of any substantive "break" with the tradition, they
are certainly right to point to a profound transformation at work in
what we have come to call the "postmodern." My wariness of the
language of the break, as well as with the term *post*modern itself, is
that both reinstantiate a "before and after" that has always been
inherent in the line as the primordial figure for time.

Derrida, who is always entangled in discussions of the postmodern, has done more than anyone to cross all the lines of transmission and communication, and—following, in particular, Freud and
Heidegger—to complicate the "before and after" by showing that the
past is always coming to meet us. In, for example, *Given Time: Counterfeit Money I*, he gives us his line, which is always seductive, tangled,
and without any hope whatsoever (thank goodness) of it ever being
straightened out.

> A discourse on life/death must occupy a certain space between
> *logos* and *gramme*, analogy and program, as well as between
> the differing senses of program and reproduction. *And since
> life is on the line, the trait that relates the logical to the graphical
> must also be working between the biological and the biographi-
> cal, the thanatological and the thantographical.* . . . What one
> calls life—the thing or object of biology and biography—does
> not stand face to face with something that would be its oppos-
> able ob-ject: death, the thanatological or the thanatographical.
> (1985, 4, 6)

Life and death are (mis)aligned, (a)symmetrical. They are not
"face to face." We cannot break the power of the line head-on; we
can just barely, if at all, face up to its implications. Death and life
abide in the very heart of the other, and letters, whether considered dead or living, are impossible without the thanatographical.

This asymmetry, in various forms, will demand our attention many times over.

Emmanuel Levinas, arriving next in line, remarks that "the 'movement' of time understood as transcendence toward the Infinity of the 'wholly other' does not temporalize itself in a linear way, does not resemble the straightforwardness of the intentional ray. Its way of signifying, marked by the *mystery* of death, makes a detour by entering into the ethical advance of the relationship to the other person" (1987, 33). Time, unlined, is interwoven with death as a mystery, a certain type of detour, and the ethics of otherness. It is not that the "intentional ray" vanishes, but, rather, that it is relativized or contextualized by another type of temporality evoked by the *trans-* or the *in-*, that, without being linear, sets linearity into motion.

And, finally, Giorgio Agamben writes in the *Coming Community* that

> the individuation of a singular existence is not a punctual fact, but a *linea generationis substantiae* that varies in every direction according to a continual gradation of growth and remission, of appropriation and impropriation. The image of the line is not gratuitous. In a line of writing the *ductus* of the hand passes continually from the common form of the letters to the particular marks that identify its singular presence. . . . (1993, 19)

"Not gratuitous": granted. But even here Agamben is speaking of the "taking-place" of anything whatsoever as a "scattering in existence" and of the move from potentiality to act as an "infinite series of modal oscillations" (18). It is not accidental that the line, oscillation, infinite series, singularity, and writing occur—take place—together.

The demonstration of the line in the history of philosophy is not a magical show, a prestidigitator's delight, but an argument, a stately demonstration of the power of rational persuasion in which reason and its reasons will be aligned in formations, marched across the page in precisely coded lines, and swerve in time to the sound of a command. Socrates—what we will call "Socrates"—begins the fundamental project of the tradition of rationality that has placed us where we are and casts us ahead of ourselves far beyond the boundaries of nuclear power, genetic transmutations, artificial intelligence, and the human capacity to create new life-forms while destroying others. Including, perhaps, our own.

We are, just now, having the experience of *being* on-line, as both subject and object, even though the lines have long been drawn that have prepared the network that puts us on-line whether we want to be or not. Technologics, as the reordering of the lines between animate and inanimate, the numbering of the real so that the virtual becomes possible, is the response to one of the oldest and most primordial of questioning calls. It is unclear, of course, who, or what, placed the call that is simulcast in our direction, but we are summoned to respond as best we can, which will always be with a falter. Heidegger wonders whether this "belonging together of summoning and hearing, which is always the same" (1958, 77) could be Being itself, but "Being" seems too staid, too burdened with history. The "summoning-hearing," however, is a dynamism that poses questions to all of us living at this busy intersection of the enigma and the automobility of what Janicaud has designated as the "powers of the rational."

There are innumerable ways to trace the call, through the tangle of lines and paperwork, through all the powerful transmitters, transcoders, and transformers, through the climate changes and tectonic shifts, and through the revolutions and revisions that have brought us to the "just now" of this boundary crossing, but three, for me, speak-up louder than the multiple others.

Ghosts.

Machines.

Cyborgs.

All are figures that have crossed over, and that assist us in thinking the crossing of the old lines between the living and the dead. All are outlaws, renegades from the proper, going back and forth by day and night, sometimes in disguise, even though the border patrols are everywhere. And all of them exist in a state resembling suspended animation, which cuts both ways, suspended between two apparently distinctive ontological realms. The dead are alive and the living are dead. *Scheintod*, the Germans say, for suspended animation. Seeming to be dead. The shining dead.

In the *Republic* Plato, through that microphone called Socrates, divides the world into a grid: "Represent them then, as it were, by a line divided into two unequal sections and cut each section again in the same ratio—the section, that is, of the visible and that of the intelligible order—and then as an expression of the ratio of their comparative clearness and obscurity you will have, as one of the sections of the visible world, images" (*Rep.* 509e).[5] On it goes: proportions and cuts, the clarity and darkness of both the soul's perceptions and the objects of knowledge, the ascent and the descent. All are diagrammed as one version of Socrates' "similitude of the sun" and Descartes is already waiting in his Dutch oven with his x's and y's. These letters burn as the line becomes a grid; the grid becomes the *Gestell*.

Philosophy begins, Plato says, with the giving of the finger. Or fingers. One, then two. The visible and the invisible; that which "summons" thought to what-is. What calls us to thinking? Discussing what is common to all the arts, sciences, and forms of thought, Socrates speaks to Glaucon of the "trifling matter of distinguishing one and two and three. I mean," he continues, "in sum, number, and calculation." He gives the example of the "smallest, the second, and the middle finger" and argues that once we are required to make not simply sensory perceptions but rational distinctions—between the one and the others, between large and small—then we are required, by the summons of the rational, to distinguish the visible from the (invisible) intelligible, and the soul is thereby "turned around toward the study of that which-is." And what Socrates names the "summoners" depends on perceptions that "strike the relevant sense at the same time as their opposites." The dialectic starts to hum once we raise our fingers. Being, through the sorting mechanism of an array of distinctions, dials us up through the fingers, calling into presence the rationality of analytic calculability. (*Rep.* 522b–25b)

Already, and philosophy has barely begun, the hand becomes severed from the mind; philosophy creates an amputated body, which will float to the surface of a pond in *The Glass Bees*. The summons, however, may not always occur as the rational—the replacement of the eye of visibility with the eye of intelligibility—but also of the oracular, the magical, the dreamlike, the phantasmatic, and maybe even (though Plato of course rejects this possibility) via the *trompe l'oeil* of the reproductive arts. After all, by what, ultimately, is Socrates called? What does "Apollo" name if not a power that traverses rationality for the sake of the rational? And how will philosophy develop an ear for the flute?

Thinking, then, as the technologics of metaphysics, depends on the capacity to make distinctions through both language, and, perhaps more fundamentally, through counting. Number is the sign of difference that must be gathered back together into the one through an ascent into a higher, supersensible realm of the sun of intelligibility. In the Platonism that has governed the broadcast of thought, there is no thinking without calculation, at first on the digits of the hand and then through the reproductive digitization of the world. And this generates a nuclear dynamism. Can you hear it? The engine of dialectics long ago purred into life and what won't it attempt to

consume, to incorporate? Can there be any residue that will be able
to resist the monopolizing force of incorporation? Will the "prefer
not" of the forlorn figure that stands at the center of what Derrida
calls that "immense text of Melville's" (Derrida and Ferraris 2002, 27)
be sufficient? And what, I will ask in "Temps," are the other figures
for the "prefer not"?

The lines of metaphysics—with number, the ideal, and all the
figures of speech—are already, as an *architexte*, in place before the
text is written, but Plato's formulation also constructs the house of
philosophy that includes a shadowy, abysmal basement—a basement
haunted by the shades of the dead and in which it is impossible to
find the foundation—and a skylight to catch the sunlight as it falls
from heaven. Turning toward and away from that sun, that radiant
hope of the good and the true, we are Plato's ghosts, his miming ven-
triloquists, his cyborgs, repeatedly drawing those lines in the sky, on
the earth, and across the palpitations, now governed by pacemakers,
of the human heart.

The heavens, the earth, and the body are all mapped by longitude
and latitude; all are precisely positioned by the Global Positioning
System that crisscrosses our bodies, whether so-called political or so-
called individual. Everything is brought under the surveillance of the
electronic panopticon and the always opened eyes of the superego of
the so-called authorities watch without blinking (although blinking,
as well as other oscillations, is something to which we will return).
Everything is recorded on camera, on tape, on memory cards, on all
sorts of externalized memory machines that are precisely timed and
dated. As our fingers, retinas, and DNA codes are scanned into the
multifoliate database that already knows how to address us, with our
preferences encoded, we will not be forgotten by the system of surveillance even if we, casting "natural" memory aside as antiquated,
forget who we are and where we have been.

All of us live, still, in that multileveled house of Platonism, even
if the destruction of that dwelling is also part of what we are now living. Maybe we will be more open to the wind and the weather, the
rags of clouds that scatter across the winter moon. And something,
perhaps a lost twin, was killed as the dialectical science of the rational was being born. The one life required the sacrifice of the other,
and great gains have been accomplished because of that offering-
victim. But as the rational culminates its deep desire to overcome
death, which is to overcome nature by reconstructing nature, the
stillborn returns as a planetary, as well as an individualized, ghost.

Whatever returns is uncanny: it disturbs and dis-places us.

It makes us toss and turn; it dreams troubling dreams.

What are its names?

Melville writes: *the unsurmised. The unaccountable.* We will, following others, speak of the *singular, the incalculable.* In a simple sentence, Baudrillard remarks that "Thought is singular, and in its singularity, thought may be able to protect us" (2000, 29). I am not sure "protect" is quite the right word, but in any case more than once, many more times than once, we will have to encounter this *un-*, for where boundaries are breached and broken the *un-* is present: as the unknown and unknowable, the unconscious, the undead, the negation of that which is. Death is one word for that negation, but, that said, we risk losing track of the line-between that we will try to follow, for death is not the opposite of life.

Heidegger gestures toward the dimension opened up by this prefix, asserting that "The metaphysics that begins with Plato within Greek thinking itself was not up to the essence of the 'negative.' Even though it escapes being equated with the empty nothing, the negative is always conceived as something negative in the sense of the lesser, something that ought not be, the *un*" (1996, 77). He reiterates this thought in his letter to Jünger: "These are questions," he insists,

> which exhibit a special sharpness while passing 'across the line,' for this passage moves in the realm of nothingness. Does nothingness vanish with the completion or at least with the overpowering of nihilism? Presumably, overcoming is only attained when instead of the appearance of negative nothingness, the essence of nothingness which was once related to "Being" can arrive and be accepted by us mortals. (1958, 79)

The transformation of our relationship to the *un-*, the uncanny within technologics, will require a transformation of language, a "new" writing that attends to the affinities between the miniscule and the gigantic. And the *un-* will be in incessant play with the *re-* and the *in-*, to which we shall also return, which shall again and again turn us toward itself.

The incalculable does not, at least not necessarily, oppose itself to the calculable. Rather, the calculable, with all its magnificent power, operates within the *in-*. Speaking about the "singular," and with Kierkegaard in the vicinity, Derrida explains in *A Taste for the Secret* that

> [T]he densest knot where this question is concerned is a knot that not only knots together each time, in a single conjuncture, the aleatory, alterity, and calculative rationality. A decision has to be prepared by reflection and knowledge, but the moment of the decision, and thus the moment of responsibility, supposes a rupture with knowledge, and therefore an opening to the incalculable—a sort of "passive" decision. In other words, one cannot rationally distribute the part that is calculable and the part that is incalculable. One has to calculate as far as possible, but the incalculable happens: it is the other, and singularity, and chance, without one's being able to do one's part; the parting between reason and its other . . . does not obey a logic of distinction, it is not a parting with two parts. (Derrida and Ferraris 2002, 61)

If it doesn't obey a "logic of distinction," which as we have seen is the foundation of the *logos* for Plato, then the structure of the logic of (non)identity and dialectics is overturned but not obliterated (as deconstruction never tires of reminding us). Whatever it is that is appearing as the transepochal logic of suspension will not divide itself again into binaries. It is an emerging configuration that combines and hybridizes what has long been thought of as distinct regions of ontology.

Calculation takes us as far as it can—and we do not yet know the limits of that being-taken—but both "before" and "after" the calculating moment stands the *in-*. The *in-*, this smallest of prefixes, almost nothing in itself, marks the boundaries of the calculable and resists—in a way similar to Bartleby's resistance—the absorption of the world into the programmaticity of the calculable. And the *in-* not only frames the calculable—the rational as the numerical, as the possibility of analysis that rests on distinguishability, and the predictable—but is within the essence of the calculable itself. Rationality, since it is so bounded, "cannot rationally distribute the part that is calculable and the part that is incalculable [and] the parting between reason and its other . . . is not a parting with two parts."

Rationality cannot, finally, distinguish itself from its others; that is, it cannot act rationally on the whole. It cannot be itself and do its proper thing. It is not the self-generated and autonomous knowledge of absolute reflection. But the "parting" that runs like a fault line through the rational, without which the rational could not be itself, does not divide into two, and more, parts. It knots. In a related, although not identical, move, Janicaud argues for a rational-

ity of what he calls "partage" or "contiguity." Unlike the rationality
of domination, it

> has no plans of conquering the world or of utilizing beings.
> Its creative provinces do not objectify themselves in any de-
> finitive method; it does not exhibit them, it leaves them
> coiled in the depths of the spirit. Its temporality is not linear;
> it does not sustain any cumulative process; it precipitates
> the cycles of an incessant, although unpredictable, Return.
> (1994, 242)

This is the rationality that carries with it its own enigma, its own
secret, accepting the encounter, within the *logos* of *technē*, not only
the majesty of number and definition, but also of a swarm of phantasms that accompany the serial movement of the rational. These
others, the ghosts of the banished, open the rational up to the incalculable, which has been the "invisible shadow of the calculable . . . since the establishment of the method as a project of the
mathematical mastery of nature and of being in general" (46). Or,
as we have learned from Novalis, whose friends in Jena are also
close at hand throughout this discussion of rationality and its others, "we always seek the absolute, but we only ever find things"
(1997, 23). The indetermined is determined.

The (in)calculable is (in)separable. Everything is distinguishable
from everything else; nothing is distinguishable from anything else. It
is the logic that stands between these, which I contend governs the
transition we are undergoing, that technologics tries to name. It is a
logic of deconstruction and reconstruction, a logic of suspension
bridges. Derrida names it, in the *Specters of Marx*, "hauntology," re-
marking that such logic "points toward thinking of the event that necessarily exceeds a binary or dialectic logic, the logic that disturbs or
opposes *effectivity* or *actuality* (either presence, empirical, living—or
not) and ideality (regulating or absolute non-presence)" (1994b, 63).

Just after the passage on the calculable, the incalculable, and the
parting of the ways of the two, Derrida admits that "*Here* is the
enigma of this situation in which I get lost" (Derrida and Ferraris
2002, 61). He is not alone. We are all lost in this enigma, in this riddle that has no single conceptual answer, but to which we must nevertheless make a multitude of responses. As we straddle the organic
and the inorganic, the machinic and the spiritual, we'll be mountain
climbing, then rappelling and spelunking, as we traverse the lines of
letters that form a virtual terrain of heights and depths. Sometimes

in the utter dark. Please watch your step, for this ascent-descent is a much-used ladder that hubbles the stars at the very edge of the universe, prepares for a quantum nanotechnology, and rewrites the human genome as the new Book of Life. Some of its rungs, having borne so much weight for so long, have become rickety.

The ladder goes up and down, bearing with it the return of repetition that is not quite a copy, not quite what the good Athenian had in mind. I would prefer not to have to mention, certainly not this early in the game, Freud's eviction from home, his fainting spells on the Acropolis, his loops through the red-light district, his definition of philosophy as paranoia, and his reading of Hoffmann's tale of the man with the bad eyes, of which there have been so many in the history of the west. But the doctor, moving us away from home on an interminable train, is always in.[6]

Welcome to Berggasse 19.

Is my collar straight?

Lie down. What comes first to mind?

Pretend I'm not here. You are alone, with your dreams of love and the death of love.

Free association that is programmed by the *un-*. *mimēsis* and repetition with a difference: the uncanny: the re-turning: the prosthesis of the time of the hour of analysis. The detour and double shuttling of *Nachträglichkeit*.[7]

It is a cliché, a word derived from the double type of the printer, something repeated so often that it has been worn down almost to the smooth meaninglessness of the Nietzschean coin, that we live in a period of crisis in which memory, truth, value, identity, and imagination are all eroding, or being shattered, as we are whirled around in the cyclotron of the dis- and en-framing that marks the end of philosophy, literature, the author, and history (among other disciplines, institutions, and concepts). We live in the end times, a time in which the powers of repetition, alongside the powers of destruction, have become unavoidably and automatically *Traum*-atic: dreamlike, mechanical, and agonizing. Here at the end of not only a millennium (and that which *measures* the millennium as millennium), the world without end is the world infinitely copied by the *Gestell*.

Xerox: beginning and ending with the *x* of a chiasma. *X* marks the spot. Of what? Treasure? The grave? Or is it a signature of one who can barely write? Who can only, nearly illegibly, mark, without really signing a proper name? Is the *X* the simple crossing-out of Being, the *kreuzweise Durchstreichung* as Heidegger phrases it, even as Being continues to appear? Or, as a complement to the negative, perhaps "the symbol of the crossed lines can [also] point into the four areas of the quadrangle (*Geviert*) and of their gathering at the point of intersection" (Heidegger 1958, 80, 82). Derrida comments that the X is a schematic diagram of the entirety of dissemination. Whatever that might mean. And: "X: not an unknown but a chiasmus. A text that is unreadable because it is *only* readable" (1981, 362).

What is a line? Where are we on the line?

Are there other ways of being than being lined up?

Technological duplication—as digitizing, cloning, and the creation of artificial intelligence[8]—undermines all notions of uniqueness, originariness and originality, individuality, and the singular. If there is nothing singular, then not even "my own" death remains mine, for there is no longer any "mineness" (although this does not at all suggest that an "ourness," in any of multiple fantasies of community such as the global village, comes to stand in the stead of what is/was mine). Being-toward-death, indefinitely deferred, will no longer be able to orient "my" life. Futurity itself, in the name of the now, may simply vanish.

And, yet, we will continue to attempt to speak of *singularity*, organized around another set of coordinates that are not coordered by that which can "fix the position of any element at any given moment—an animal in a game reserve, a man in a business (electronic tagging)" (Deleuze 1995, 181). The bio-scopic-electronic-regime that is being constructed is undoing the lines of demarcation between matter-earth-animal-human-thing and is constructing a new type of being that can be abbreviated as a cyborg, which is a hermetic-hermeneutic figure that straddles all the old borders.

Once upon a time—and we all know this story—there was a line. The world began; we are now in the middle, or perhaps just past the middle, of the tale; and, finally in one great blaze of light, the world will end. Gregory Ulmer comments,

> Grammatology confronts nothing less than the sediment of four thousand years of the history of language, during which time everything that resisted linearization was suppressed. . . . The linear schema of unfolding presence, where the line relates the final presence to the originary presence according to the straight line or circle, became a *model*, Derrida says, and as such became inaccessible and invisible. Given Heidegger's demonstration that this mundane concept of temporality (homogeneous, dominated by the form of the now and the ideal of continuous movement,

straight or circular) is the determining concept of all ontology from Aristotle to Hegel, and the assumption that the
linearity of language entails just this concept of time, Derrida concludes that "the meditation upon writing and the
deconstruction of the history of philosophy become inseparable." (1985, 8)

Now that the deconstruction that we might call literature, a weath-
ering that has always surged about within the normalizing male
body of philosophy, has itself become far more visible since a number of writers have injected dye into its bloodstream for magnetic
resonance imaging purposes, the line of ontology has become
frayed. Literature has once again surfaced—along with psychoanalysis, another discourse of repression—and, in order to follow
the line from Plato to the emergence of the posthuman, we will have
to read across all those old generic lines of division that separated
the discourse of truth from that of the errors of the imagination and
the phantasmic. This, perhaps, will produce another temporality, at
least as a complement to the line of cause and effect, and another
range of possibilities for the thought of the (post)human.

What it means to be a human being—although "meaning" itself
is at stake in the reordering of the front line, the sidelines, and the
end line—will no longer have its locus in any of the many varieties of
tradition, of the so-called community or in the post-Enlightenment
individuality of the liberal subject (although we may well want to
rearticulate some of these values along other lines), but in the
(non)system of the synthesized machinic-organic. In other words,
human identity, as the fantasy of an identity not in symbiosis with
the automaton, will be radically modified. Part of our task is to re-
fantasize the human and to project this image, the many images of
plural imaginings, into the future as best we can so that the image of
the posthuman is not completely fixed by the powers represented by,
for example, the lawyers on Wall Street or by Signor Zapparoni, the
monopolistic master of duplication in *The Glass Bees*.

If it is the case that the autonomous liberal subject is vanishing
and that there is nothing but the repetition of the always already repeated of the socially distributed intelligence of the automaton, then
we are all only listening to the reverberations of echoes from what
used to be called the past and watching copies become not a virtual
world, a version of the primary world, but the world itself. As Baudrillard has put it, our time is a period of the "vanishing of actual history, a vanishing of the event in the information space. This amounts

to making the past itself into a clone, an artificial double, and freezing it in a sham exactitude that will never actually do it justice" (2000, 40). The so-called present is also being cloned in reproductive factories like Las Vegas and television shows that deal with "reality" either in a supposedly direct way or by incorporating current events into the televisual scenario. The "secondary" replaces the "primary"; the copy replaces the original, and we encounter, in Baudrillard's already well-worn phrase, the "precession of the simulacra." The ordering ordered by technology has its social form, and it is violence. The true "nuclear fallout . . . is the meticulous operation of technology [that] serves as a model for the meticulous operation of the social" in the form of the "terrorist rationalization of the social" (1994, 35, 37).

Cloning—like capitalism, democracy, metaphysics, and all the other simultaneous epochs—entails a temporality, a form of identity, a relationship with the natural, and a question of how, if one can, to "do justice" to the situation. If technological duplication "freezes" everything, even as it simultaneously sends everything into a mad frenzy, then that may be the end of any talk of justice. All the ancient lines are being crossed and we are being crisscrossed. To the extent that I am able, I want to look both ways before I cross these lines, which is never of course a simple and singular line, but a twined cable, fiber optics running off in all directions. These tracks, disappearing into infinity, however understood, are extremely dangerous and are never empty. Bullet trains are always bearing down and arrive before our eyes, ears, hands, or hearts have time to react. A train is always leaving the station in Vienna. Everything shakes and trembles. We are outpaced by the productions we have produced, and must learn both mourning and exhilaration, if that is possible.

Ghosts and cyborgs: phantasmatic echoes fill the chamber of my inner ear, which is, admittedly, sometimes a tin ear. I hear what I hear; I leave it to you to sort out what is important and what is not.

What machine will you use to do the sorting? Do you know? What kind of machine are you? What voice emerges from this singular gathering of written voices read and (re)written, from the living and the dead, is *yet to be*—

The line is dead; the line buzzes with life.

Go ahead.

Memorize your lines; you're becoming someone else.

Reading line by line, traverse the lines

3

The Platonic
Teleport

Who knows if being alive is really being dead, and being
dead is being alive?

    —Plato, quoting Euripides (from an unknown source)

Cyberspace is Platonism as a working product.

            —Heim

Plato launches metaphysics, the thought-program that will give rise to cloning, artificial intelligence, and the crisscrossing of the (in)animate at the turning of the transepochal. And this Platonic template—with multiple revisions, to be sure—is teleported as far into the future as we can see. The file carries the name, among others, of cognitive science, the human genome project, AI, and virtual reality, and contains within itself, although there is no real "within," instructions on how to overcome death and become immortal. The process will involve the construction of powerful prostheses of memory, organ harvesting and manufacture, genetic codes, and the exchange of what used to be called the "natural" and the "artificial."

The planet is becoming mechanic-organic, biotechnical, and the Platonic software insures that *bios* and *technē* will couple and breed. The experience of time itself is, again, warping, bending, and folding in unexpected ways.

Teleportation is itself a product of the research project launched by Plato. An article in *Nature* on the teleportation of photons—and it is entirely appropriate that light is the first to make the switch—opens as follows:

> The dream of teleportation is to be able to travel by simply reappearing at some distant location. An object to be teleported can be fully characterized by its properties, which in classical physics can be determined by measurement. . . . But how precisely can this be a true copy of the original? What happens to the individual quantum properties, which according to Heisenberg's uncertainty principle, cannot be measured with arbitrary precision? (Bauwmeester et al. 1997, 575)

The answer has something to do with what physicists call "entanglement," which occurs when measurements on one particle change the properties of another in exactly the same manner. This produces what Einstein called "spooky effects at a distance." In the most recent discourse of quantum mechanics, the discourse markers of technologics are all present: the dream, measurement, copies and originals, a structure of entanglement, and, naturally, the spookiness of the whole thing. Physics, philosophy, literature, and psychoanalysis all seem linked in the same entangled web of what used to be called the *Zeitgeist*, time's ghostly perturbations that echo throughout the sociality of language.

The master code of the Platonic program that allows us to think teleportation and all its affiliates in telecommunications and the biosciences is the idea of *mimēsis*, necessary for both the reproduction of the same and for Socrates' immortality as he makes the crossing, *psyche* leaving *soma* behind, from Athens to Hades. Language, memory, consciousness, communiques, culture, and all the rest require a concept of re-production and copying, faxing forward to a destination more or less preestablished by the choice of code and operating systems. Such an operation is not, of course, the "imitation" involved with sophistics and poetry, which are supposedly banned from the premises, but rather a signal sent through the transmitting towers, glinting in the sun, of dialectics. But to discuss, for the n+1 time, the problematics of *mimēsis*: how melancholy, how mournful. A now-grizzled Narcissus scraping his slow way, with bloodied knees, to the pond yet again and leaning over to gaze in the wrinkles of the water at his pale, dead face. What is it, at the end, that links repetition and death? How can we understand another repetition that rephrases the lines that order the difference between death and life?

And yet, for what I take to be a very short moment, a moment already vanishing, we have the slightest of advantages and "today's process of transition allows us to perceive what we are losing and what we are gaining—this perception will become impossible the moment we fully embrace, and feel fully at home in, the new technologies. In short, we have the privilege of occupying the place of 'vanishing mediators'" (i ek 1997, 130). We exist as nanotechnology, and its culturally parallel ports are emerging and are able, with the speed of a glance, to look backward and forward from a different cog on the unhinged gears of the history of thought. There is a history of philosophy and philosophy's adversarially complementary others, literature and psychoanalysis, disciplines that look at the thing—including the human thing—and "see it slant." And the complex network of (non)philosophy that includes Nietzsche, Heidegger, and Derrida is working diligently, to unthink metaphysics and to learn to think otherwise. Since this networking, which we can abbreviate as deconstruction, is not a program, or, at least, because it positions itself at the margins of the program in a scribble rather than in a normative well-structured typeface, it is not predictable.

Not, at any rate, completely.

There's a ringing in my ears. There's a call from Athens: again, the program is relooping, coming again around the bend of the bay. Call Socrates back from the dead (for Greece, dial 30; the city code for Athens is, what else, 1; then punch in the person-to-person extension for Hades that is universally accessible). He's muttering something in the dark. Let's ventriloquize him, as he ventriloquizes us.

This call may be monitored to ensure quality-control.

A listening post, an echelon of listeners—a person, a machine, a god, a government, a conscience—is always set to overhear and record our conversations.

WIRED METAPHYSICS

Philosophy, wired from the beginning, speaks with the voices of ghosts. The lines, crisscrossed and entangled, buzz with gossip and tall tales, with rumors of gods, grounds, goods, the life beyond life. Everything is a long-distance call that originates in the telephony of our inner ear, always historicized. But it's pandemonium; everyone is talking at once. If we're going to make anything out of all this chaos we'll have to learn to count differently, but that will come, if at all, only with time. Listen! It's ringing. You've got him on the line. He's answering, responding to your call. He's back from the dead and phantasmatic letters rise from the grave of history.

He's speaking to us through a prerecorded voicemail program. Since his voice—and what range, what timbre and bravura!— sounds through an already technologized speaking system, we'll have to acknowledge our already developed technologized ear in order to hear anything of value. We're already a psychosocial apparatus of bundled prosthetic networks that constantly talk back and forth, refer to their own memory functions, and incessantly translate one "level" of discourse to another.

It's uncanny, this *daimonic* telecommunications system.

Plato labored long hours to give birth, through a midwife, to a transcendental fiber-optic network, and he calls it *eros*, he calls it dialectics, he calls it the love of wisdom. It is both dead and alive, always wanted and wanting. Its logic is off the grid of the calculation of value, whether in ethics or economics, that it itself sets in place. This place, this *epistēmē* that is a table of values, is always moving. The ghosts it generates, and is generated by, are restless, unable to stay put on either side of the line that supposedly divides the dead from the living, the phantasm from the thing, truth from opinion, value from its others. They return to set the house in order, but what if some business can never be completed? Credit not given; debts not paid?

What if *haunting*, a structure of spectrality (and what I will return to as phantomenology), is an ontological mode that is constitutive of the ghostly missives of Being itself? An unsettling experience that we will never be rid of, unless the posthuman regime finds a way to destroy the question itself, and all the detours of signification necessary to the asking of the question.

Plato constructs the power grid, the systems of transformers, the network of transepochal lines that have governed philosophy, and, finally, the possibilities that are articulated by cybernetics, the command-and-control system that steers us beyond the millennium. This power grid is called rationality, which Plato sets up as a determined use of language—that is called, for example, "philosophy" rather than "sophistry"—that proceeds toward the abstract universal through the fields of logic and mathematics. This grid establishes the dividing lines between illusion and lucidity, between shadows and light, between the immortality of *psyche* and the mortality of *soma*. If one is bitten by the snake, stung by the bee, then one climbs through the grid, keeping the eyes shaded in order to see against the glare, up and through the technological scaffolding of the dialectic.

Brightness.

The earth recedes.

The curvature of the horizon of being glimmers as if there were a beyond.

The days of this grid are numbered, have always been numbered. Or lettered: oriented to the points at which x crosses y. We are plotting this graph as it disintegrates—as it has also always been doing—and sets itself into an inscrutable motion that will not be programmable on any graphic calculator.

This motion, incalculable, that shifts the shape of the grid in its entirety is what I am attempting to descry, to read with my hands while crawling along through the dark. I am, as it were, reading backward from this juncture in time, all the while listening with turned head for faint sounds of the future. There is, among these arabesques of reading, a certain virtuality produced by the machinations of that which has been divided and designated as the natural-artificial. It has been called the animate and inanimate, natural and mechanical, the living and the dead, the soul and the body. As if one could exist without the other, be distinguished, in any rigorous sense, from the other. As one commentator on Derrida's *Specters* has said, it is the "cyberised systems that make openness to the other, and to the event, a more difficult and more urgent task in the face of the cata-strophic beat of tele-technologies. In is that very 'system,' however, that ensures the apo-strophe will reach us, however hauntingly" (Luckhurst 1996, 182).

We have always been technological, technologized, but with the appearance of the technologies of computing, cloning, engineering at every level of the body's organization, and of systematic surveillance, ancient dreams are becoming instantiated as everyday life, and, as such, threatening to recede immediately into the invisibility of habit. Thinking must resist this recession, and writing must continue to act as a dye, an X-ray, an MRI processing that brings the usually invisible into the range of readability. If we cannot read, if there is no time for reading, ethics will have been evaporated by the heat of a nuclear blast, swept away by the solar wind.

TELECOMMUNICATION WITH THE DEAD

Ghosts, they say, are abroad at night and sleep during the day, whatever "sleep" might mean to those spirits who are dead and alive. Logic, with its basis in the law of noncontradiction, is not a salient concern for ghosts, except for that "other logic" of revenge that has supposedly been superseded by the institutionalization of the rational and its formulation as law. But then again, Socrates, in the *Apology*, looked forward after cheerfully quaffing the hemlock to long discussions with the heroes and poets waiting for him in Hades. Death, to his imagination that had been tutored by "what we have been told," was not a place of answers to the question posed by existence or of some final revelation of life's meaning, but rather a locus for further questioning. The rational, then, must be more encompassing than the realm of death itself. Truth, falsehood, language, reason, time, and all the other necessary transcendental conditions for asking questions must, for Socrates, intersect life and death, Earth and Hades. This is the basis for the transcendental hope that technology will exceed death, limit it, and bury it for good in the crypt of the past. Life will continue as life, unbroken by the breakwater of finitude.

However, the dead (at least, the famous dead that Socrates mentions) know *more* than the living—they know their own form of existence as shades—but they do not, apparently, know the truth of all things, for they are open to the joys and embarrassments of a Socratic examination, for which the philosopher, that time around, will not be liable to execution. How, then, will the uneasy and anxious dead rid themselves of Socrates? What would the contours of this conversation among the shades be like? Surely the heroic dead would not simply reminisce about the good old days and regale each other with stories of severed heads, battle frenzy, and sword-wielding women? Such would not interest Socrates in the least, since poetry in all its forms has been superseded by philosophy.

His will be the voice of the rational entering the domain of its others: the irrational, the transcendent, the night, the poets. A bit giddy about the possible chats with the glorious dead, he notes that he will compare experiences with those convicted unjustly, such as Palamedes and Ajax, but, more importantly, he could spend his time "testing and examining people there, as I do here, as to who among them is wise, and who thinks he is, but is not" (*Apol.* 41b). Death itself, apparently, does not destroy self-deception. The *daimonic* voice of Apollo, invited or not, takes up residence in the halls of Hades, reason launches its millennial probe of death itself. Even bodiless, it keeps speaking. (William Gibson calls this a "construct" and names him "Finn," again.)

Socrates has already argued that either death is nothing or a "relocation of the soul from here to another place" (*Apol.* 40c). If death is nothing, then there is nothing more to say. If, on the other hand, death is a transmigration across rivers to a flowery meadow, full of honey bees droning from one bud to another, then questions arise. Socrates declares that the dead "are happier there than we are here . . . and for the rest of time they are deathless, if indeed what we are told is true" (*Apol.* 41c). Approaching the enigmatic line of death, Socrates turns back to "what we are told"—a folding back of rational critique into the traditional, probably Orphic, narrative of an afterlife—and to his own personal "mantic sign," which has not said its "no" during the course of the trial, thus giving Socrates "convincing proof" (*Apol.* 40c) that he is embarked on the correct course and that death is not an evil. There has been no no— and this must be read as a yes. (Francis Picabia has given this definition of Yes: "Yes=No." Freud has said every no is a disguised yes. Bartleby says no, as, perhaps, a form of yes; Nietzsche does the same and then just says yes. Derrida has said that deconstruction is an affirmation. This learning how, and when, to say yes/no is the movement of ethics.)

This is not, of course, true knowledge, *epistēmē*, in a strict sense, for it is not rationally justifiable, it is not the conclusion of an argument. Socrates cannot *explain*, logically, either the old stories or the existence of his own sign, for they both come under the rubric of "this is what we, or I, have heard." A voice, as it were, from beyond. And yet both are essential "grounds" for the dialectical process that so engages him. They are, as it were, the unconditioned, which while not giving the logical *grounds* for his questioning—that is, after all, the goal, not the beginning, of the Socratic process—they do give inquiry a boundary to the field in which the rational will labor: tradition and the word of the god to the singular listener. The rational will mark the destruction of such traditions, but it will also form a tradition of its own, give itself over to its own keeping until, for the most part, the voice of the god will fall silent. Outside the mouth, ear, and the hand of the rational and its modernities.

But Socrates, is a liminal figure, living on the boundaries, and *singularity* cannot exist except in relation to the *repeatable*, which Plato calls *mimēsis* and Derrida "iterability" (which are [not] the same). We must, repeatedly, sound out the echo of "Socrates" as he is produced by the engravings in the acoustic tracks of recorded history. And with this structure of singularity-repeatability, we are *already* on the verge of cloning, not through that which we, today, call technology, but in the oneirically metaphysical register of thought from which the technological, as instantiation, will emerge.

This thought, which Derrida has taught us to understand as the interweavings of the structures of textuality, will of course pass through an immense circuit of complexity as the sciences develop; but, nonetheless, Socrates-Plato designs, with its machinery of dividing lines, the template for biotechnology and technobiology. Socrates, represented after his death by the voice-activated writing machine of his student, sends his soul toward timelessness (although this is only a metaphor, as his speech with the dead will, presumably, require time),[1] and now we can imagine, if not yet, not quite yet, accomplish, electronically forwarding our "selves" infinitely into the future. As Hans Moravec has hypothesized, human memory will eventually be codable and thus transmittable ad infinitum into the future. More generally, he notes that

> As the machinery grows in flexibility and initiative, this association between humans and machines will be more proper as a partnership. In time, the relationship will become much more intimate, a symbiosis where the boundary between the "natural" and the "artificial" partner is no longer evident. (1988, 75)

The transepochal is the sign of this ancient, imminent "intimacy," even if such a relationship does not take the form of electronic immortality that Moravec envisions.

Socrates is not, at this crossroads, a purveyor of explanation as a philologist, a comparative mythologist, or a psychoanalyst: he is a man on the verge of death. This inability to explain would apparently place Socrates in the same position as the poets, craftsmen, and politicians that he had earlier in the dialogue examined and found wanting, but the essential distinction remains: Socrates, even in his extreme situation, knows that he does not know, but he nonetheless has a "good hope," a hope based on that which he has been told and upon his own deepest intuitions, upon that still, small voice of otherness that speaks from a within. One can only presume that he would still be pursuing the definition of the good life, the securing of the meaning of words, the *logos*, and the radiance that grants the world's intelligibility.

Reason, shaped by what we might call a visionary imagination, makes that journey again in the *Phaedo*, in which Socrates first works to prove by rational argumentation the soul's immortality, and then, at the conclusion of the dialogue, offers fables of the earth's structure and that of the afterlife in order to console himself and others, as well as, perhaps, to initiate his friends into his vision of the unity of life, death, and the world. His friend Crito begs him to hold onto each moment and extend those moments as long as possible: "The sun is still upon the mountains," he says, with poignant desperation. "It has not gone down yet. Besides, I know that in other cases people have dinner and enjoy their wine, and sometimes the company of those whom they love, long after they receive the warning, and only drink the poison quite late at night. No need to hurry. There is still plenty of time" (*Phaedo* 116e). But Socrates knows what the tradition will forget, that time is not an extension of moments along a line; it is an opportunity to accomplish one's own fate. *Kairos* supersedes *chronos*. Thus, he accepts the hemlock and reminds Crito to remember the offering to Asclepius. He is on the verge of being healed.

Philosophy, at that paradoxical moment, would no longer be a preparation for death, but a preparation to step out of death into the light of the sun called life (which appears, from our vantage point, as the disappearance called death). The dialectic, especially as it is expanded by Hegel, includes within itself this expectation and this machinery of reversal. Like a shuttle on a loom, thought would move back and forth, rendering the line of death finally meaningless. This, however, assumes life/death to be the fundamentally organizing opposition, and perhaps these two occur within a larger, more encompassing pattern that we are not able to re-cognize through conceptuality (but which nonetheless has a history of names).

Once Socrates crosses the line, however, he would be thoroughly dead and not a ghost, for a shade, after all, is not the same as a ghost that returns to haunt the old place, the old family, even if this particular shade is garrulously alive in Hades. Socrates goes to haunt the shades themselves—the dead will be haunted by the voice of one of their own. Rationality contains within itself a principle of haunting, and, having no unfinished business in Athens for which he needs to seek revenge, Socrates departs freely and will not return, at least not in the same phantasmatic mode as, for example, Clytemnestra or Agamemnon (and Iphigenia?)—to haunt the polis. One name for this genre of the phantasmic is literature.

And yet the *name* "Socrates," at any rate, *does* continue to return, to haunt Athens, philosophy, and other ruined houses with a compulsion that borders on vengeance. We can't seem to shake him loose. "Socrates" represents many characteristics—the rational, the wise, the ethical, the martyr, the decadent—but he prophesies Athens's inability to destroy "him." He remarks that the Athenians want to kill him in order to be rid of a nuisance, but, instead, his absence will leave a gap that others, critics of the polis or the cosmopolis, will automatically come to fill. And many of those latecomers will bear the name "Socrates" on their standards.

The same is true about the empty place-holder, and the place-holder of emptiness, of "Socrates" as metaphysician. When Chaerephon heard from the oracle at Delphi that his friend was the wisest man in Athens, Socrates, faced with a proposition that must, given its source, be meaningful but that was nonetheless not yet understandable, went in search of the truth of this statement. He "puzzled" over the "hidden meaning" for a bit, then went to empirically "check the truth" by interviewing the politicians, poets, and skilled craftsmen. Socrates did not, in other words, find the truth in an interior intuition or in a self-evident claim of reason, so he turned to the empirical world to see if the truth appeared, was produced, as the result of conversational interaction with interlocuters.

The truth, in the negativity of the Socratic dialectic, exists as a configuration of the in-between: truth is *already* crisscrossed by the crevasse of rational/empirical and the here/not-here (which is both a spatial and a temporal image). The truth, as a thing that can be "positively" known by measurement, for example, *does not appear.* Socrates' method produces only a blank space where truth, were it present, would present itself. Socrates concludes his quest for the interpretation of the oracle's words by noting that his investigation has created hostility—there is always a mood, Heidegger's *Stimmung*, that accompanies questioning—and by observing that most likely

> real wisdom is the property of the god, and this oracle is his way of telling us that human wisdom has little or no value. It seems to me that he is not referring literally to Socrates, but has merely taken my name as an example, as if he would say to us "The wisest of you men is he who has realized, like Socrates, that in respect of wisdom he is really worthless." (*Apol.* 3a–b)

Socrates is speaking in the confident, but not arrogant, mode of "as-if" and "most likely," and the fundamental line between the gods and mortals, *athanatoi* and *thanatoi*, is kept firmly in place. Apparently wisdom, like deathlessness, belongs only to the divine. And yet, Socrates *does* cross the boundary between life and death. In his apology, Socrates takes on, temporarily and provisionally of course, the voice of Apollo himself: "The wisest of you men is he who has realized, like Socrates, that in respect of wisdom he is really worthless.'" Apollo, as it were, takes Socrates' name "as an example," the example par excellence. It is a particular name, a name like any other, but it comes to serve as an example that at least points in the direction of the universal name. And there is, once again, the work of a certain boundary-crossing in this logic of the example.

First, the mortal voice takes on the overtones of the divine voice. Socrates is, to be sure, playing—in all senses of the word—but it is, as usual, serious play. He is acting like Apollo; or he is pretending that Apollo has donned the microphoned mask called Socrates. A dramatic ventriloquism, essentially comic, brings Socrates—with his masks, his miming, his poses, his string-pulling—close to the art of puppetry. Another line is crossed when the highest of values, wisdom, is aligned with "worthlessness." The low becomes the high; the high becomes the low. To know is to know-not, and already (and the game of philosophy has hardly begun) Platonism shows itself as a (k)not-philosophy. But the history of the West, in the metaphysical tradition that becomes technologics, will work without rest to untie this series of knots and pull the string straight.

Platonism is, in its main thrust, the philosophy of ascent toward abstraction. It moves out of the damp darkness with a fire at its center, up the divided line and the ladder of love, and away from the light and shadow play of earth and thing, earthly things. As such an ascent, that divides itself into sections, it will provide the underlying grid of intelligibility, the opening of thought in which rationally conscious cognition is predominant. And, in the long run, it will provide the grid for the development of the cybernetic space in which the intellectual imagination knows the world as digitizable, a complex of 0s and 1s, and thus can dream a technique to create artificial intelligence.

And, finally, in this passage the line between the particular and the universal, the empirical and the rational, is crossed, entangled. And not for the last time, for it is out of this net of crossed lines, perhaps more than any other, that the history of philosophy produces itself, makes itself into a house divided into "mind" and "body," "rationalism" and "empiricism," representation" and "world," with their multiple implications and star-crossed loves. The as yet unborn voices of future philosophers all resound in the interstices of this simple crossing: "It seems to me that he [the god] is not referring literally to Socrates, but has merely taken my name as an example." Able to continue serving Apollo's voice only because he is Plato's masked mouthpiece, Socrates becomes an example. He is both ventriloquist and dummy, a kind of prototype for the talking machine, the speaking automaton, a voice-activated writing program (although voice, too, will eventually be come to be understood as a kind of *écriture*).

No longer someone governed by the intimacies of empirical history and faced with the inevitability of death at the closure of "biological duration," he has become a concept that has the potential to gather to itself particular representative instances. Philosophy has left the ground, to which it will always (and always unsuccessfully) attempt to return as a conceptually defined metaphysical, or scientific, resting place. Metaphysics is a constructed space of intelligibility, and, as such, cocreates a temporality, a place of memory, an archive. Derrida calls this process "consignation," explaining that

> Consignation aims to coordinate a single corpus, in a system
> of synchrony in which all the elements articulate the unity of
> an ideal configuration. In an archive, there should not be any
> absolute dissociation, any heterogeneity or *secret* which could
> separate, or partition, in an absolute manner. (1996, 3)

This is one of the fundamental desires of technologics as metaphysics: to unveil nature, to turn the hidden out into the light of understanding so that there will be a single space of intelligibility— without secrets or heteroglossalalias—that can be perused, and used, by the drive toward the rational. But one cannot rest in metaphysics except as in a *Friedhof,* that place of contentment inhabited by the dead—and even there only if the dead are not restless and eager to continue to speak.

Philosophy is a colloquium of ghosts who continue to raise essential questions from the open crypt of the site of death. As ghosts, they are puppetmasters pulling *our* strings, ventriloquizing *us*, making us strut across the narrow boards of our generationality as we repeat their words. And while we come to life in writing and speech, we do so in a way that has mechanical aspects to it. We reproduce their texts, but, for the most part, along lines of interpretation long ago laid down and in all the conventions given by the language in which we write. We produce and are produced by what we produce. As Derrida, in *The Post Card*, remarked about the act of reading,

> Two logics, therefore, in effect incalculable [and we will return to this incalculability], two repetitions that do not oppose each other any more than they repeat each other identically, and which, if they repeat each other, echo the duplicity that constitutes all repetition: it is only when one takes into "account" this incalculable double-bind of repetition . . . that one has a chance of reading the unreadable text. . . . 1987, 352)[2]

The *incalculable* double bind: the duplicity of duplication, accounting for the incalculable, and reading the unreadable text—X—the duplication of duplicity, the incalculability of the accountable, and the unreadability of the readable text. This is a description of precisely the double-bind, strangling and liberating, in which the (un)natural logic of (non)contradiction in which metaphysics, with all of its fecund—including feces, factory, and fiction—terminology, has always been snared.

Socrates is translated into an example; the transmigration has already begun to the other side, the other place, the place of the other. All in a single line, a thread in the Platonic webwork. As an example, Socrates as a man vanishes and is replaced by a word-concept, a structure of universality-particularity that is unavoidable in the thinking of the form of rational explanation called philosophy. And yet, just as the concept is necessary for thought, the trace of the man remains as a remainder in the textual projection. After all, the example given is "Socrates" and not "Thrasymachus," "Callicles," "Meletus," or even "Plato," although these, too, become examples in other contexts. The specificity of signifiers, even though constituted within a medium of infinite substitution, matters. (This might be called *style*.)

Socrates becomes a reference point for the history of philosophy (which is already a massive interpretation), a citation system, a representative case; but Socrates as a signifier that represents is not completely emptied of content, even of important traces of that nonconceptual being, pudgy and patient, that we can only evoke with a necessary futility, who relentlessly badgered the beleaguered city of Athens. Socrates is a *determined* case and represents only certain characteristics, not all characteristics. The concept as representative cannot exist on its own; it does not flutter heavenward as the pure spirit of the idea, but is always what Freud might call *anaclitic*, always supported by its multiple others that, as it were, drag it back down to earth. The concept is always lettered, and, as such, leans. It needs support.

Derrida has gone the farthest in analyzing this supplementarity, this multiplicity within the name: singularity/universality. "It would be necessary," he observes, "to recognize both the typical or recurring form and the inexhaustive singularization—without which there will never be any event, decision, responsibility, ethics, or politics" (1992b, 80). Socrates is both "recurring form" and "inexhaustible singularization" (as is that most forlorn of wights, Bartleby, who is waiting patiently for our arrival in his chambered cubicle on Wall Street). There can be no recurrence, no repetition or duplication, without singularization; and, there can be no singularization without the repetition of the structure of the name, of conceptuality.[3] In any and every case, the law of repetition is universal; if one is to survive, then one must in one way or another be, if not cloned, then at least repeated in difference, in what Derrida will call "iterability."

Plato, even as he was miming Socrates and working to net the monstrous into the city walls of the rational, also kept a line open, at least some of the time, to the so-called other side. "It seems to me that he [the god] is not referring literally to Socrates, but has merely taken my name as an example." The god, in the form of a *daimonic* voice which always breaches the internal/external divide, has granted Socrates the name of "philosopher," and, in return, takes Socrates' individual name back as an example of the modesty of the wise. The name has been peeled off, meta-phorized away from its carrier, the previous empirical bearer of the name, a necessary but not sufficient generator of futural meaning, and becomes a representative of an unknown number of others who will follow in the footsteps of the one whose eyes point toward the sun.

In Plato's corpus, Socrates is never the "literal" as the biological-empirical Socrates—how could he be?—but is always the "literal" as the textual-figure netted by the order of letters: the name as a complex structure that whispers through libraries, universities, and throughout the Internet. (Check and see for yourself.)

At the very least, to be dead means that no profit or deficit, no good or evil whether calculated or not, can ever return again to the bearer of the name. Only the name can inherit, and this is why the name, to be distinguished from the bearer is always and a priori a dead man's name, a name of death. What returns to the name never returns to the living. Nothing ever comes back to the living. (Derrida 1985, 7)

Once again, and not for the last time, we run up against the complex of name, return, death, and (in)calculability. Socrates is embodied only by being disembodied by the fatal but revivifying translation into writing of Plato's interpretation that projects "Socrates" ahead as a ghost who will haunt the future of any philosophy, for as long as philosophy survives as philosophy. After that, who knows?

The reason exemplified by Socrates is a nascent, although extraordinarily sophisticated, technologics. The dialectic is the *technē* of the *logos*, the operative thinking machine that ratchets the philosopher away from the visible—that which appears first of all and most of the time—toward the invisible, a task that requires both erotic and sacrificial work. Technology, in this fundamental sense, does not, except as an aftereffect, create any appliances *in* the world, but moves toward a *beyond* to find its telos. This relation between "in" and "beyond" is, however, far from simple. Each term assumes that the world is somehow a container, a topography with an edge, a boundary. Both are spatial metaphors, and Plato, in addition to world-space, is concerned with world-time. When one exits the cave, one indeed does move about in a freer, more illumined space; but, also, as psyche, one moves beyond the world-time of the fatality of finitude. If, that is, one becomes Socrates, the master of dialectics, and learns to scale the crags of reason, supported by titanium pitons.

Technology, in this primordial sense of thought thinking, is always already an overcoming of space and time. (And we already hear a faint echo, before the fact, of Nietzsche's *überwinden*, especially of the madman's interrogative lamentation about who has wiped away the horizon and destroyed the "above" and "below" of the intellectual-moral order.) Thus, whenever technologics goes to work (and where does it not?), there will be a will to power, a disrupting of directionality, a certain type of rage, an entangling of space-time, and the appearance of ghosts.[4]

The Platonic order of *mimēsis*, by launching a dialectical thinking that ascends toward truth by making distinctions in definitions—and what is more distinct than the difference between 0 and 1?—provides the "initial" conceptual conditions for the digitization that marks the highest of high technology. Plato aligns the intellective-mathematical—which is also a radically displaced form of the oneiric that conceals a wish within its structure—with the ontologically privileged "highest" and attempts to think how the highest produces the play of simulacra within the "lowest" strata of being. For Plato, human existence occurs as a kind of video game, even though he labored mightily to build a ladder, constructing each rung as he went, that rose from the lights and action on the game screen into the hidden region that *produces* the shadow play taken for life.

Human existence is inhabited by the machine, in all its sensory and logical formats, from (before) the beginning. As Derrida puts it:

> Any living being in fact undoes the opposition between *phusis* and *technē* As a self-relation, as activity and reactivity, as differential force, and repetition, life is always already inhabited by technicisation. The relation between *phusis* and technics is not an opposition; from the very first there is instrumentalisation. The term "instrument" is inappropriate in the context of originary technicity. Whatever, prosthetic strategy of repetition inhabits the very movement of life: life is a process of self-replacement, the handing-down of life is a *mechanikē*, a form of technics. Not only, then, is technics not in opposition to life, it also haunts it from the very beginning. (1994a, 50)

Plato lays out the schematic grid of value—he outlines the lines that will grid the heavens and the earth—that enables the creation of technoscience. He sets up the assumptions of the command-and-control of the passions and of nature, of the ordered and knowable nature of the entirety of the cosmos, the mathematical ground of nature, and the methodological patience that such work requires. He provides a rudder (*cyber*) for knowledge, and it is *mimēsis*, the good form of dialectical philosophy rather than the bad copying entailed by both poetry and sophistics, that guides the operation of the journey of knowledge a showdown with death.

Good *mimēsis* is analogous to the desire of technologics for the perfect copy, the noise-free reproduction that we call cloning, a possibility that Plato had already projected.[5] The *Cratylus*, which is an extended, often ironic meditation on "the truth or correctness of names" (*Crat.* 384a), includes a famous example of hyper-*mimēsis*. Cratylus and Socrates are talking about whether, when a single element of a number or name is changed, the "general character" (the idea, the meaning) is changed as well. Socrates admits that this is the case with numbers, but denies it to be true when it comes to the "qualitative."

I should say rather that the image, if expressing in every point the entire reality, would no longer be an image. Let us suppose the existence of two objects: one of them shall be Cratylus, and the other the image of Cratylus; and we will suppose, further, that some God makes not only a representation such as a painter would make of your outward form and color, but also creates an inward organization like yours, having the same warmth and softness; and into this infuses motion, and soul, and mind, such as you have, and in a word copies all your qualities, and places them by you in another form; would you say that this was Cratylus and the image of Cratylus, or that there were two Cratyluses? (*Crat.* 432a)

Cratylus, of course, admits that there are "two" Cratyluses (so the word acts as both name and object of reference for the name). An exact replica of another object is a clone, and cloning has already occurred in the Platonic text, the oeuvre that teleports itself toward, among a multitude of other sites, a sheepfold in Scotland. Clones are, apparently, examples of perfect *mimēsis* at work: self-reproduction without difference or residue. The idea(l) that transmits itself, without interference, with clarity and distinction. This is the dream of auto- and heterogenesis as the same act of identity reproduction. When I produce the other it will be myself all over again.

An "image," for Socrates, entails a mark of readable difference, a "falling off" from the original entity such as the "real" Cratylus. Image formation involves a "bad" *mimēsis*, while the perfect clone would be an example of "good" *mimēsis* in that it reproduces the prior without flaw. An image is a move away from the *eidos* that occurs, even in the "qualitative," if too many "letters" are changed. When, for example, does Socrates, as name and idea, become not-Socrates? Socrate? Scrates? Cra? What is the limit of intelligibility?

Numbers, here as elsewhere in Plato, seem a more straightforward case, a purified language. The question of names and essences—the question of representation—causes some concern for Socrates. "Do you not perceive," he asks, "that images are very far from having qualities which are the exact counterpart of the realities which they represent? . . . But then how ridiculous would be the effect of names on things, if they were exactly the same with them! For they would be the doubles of them, and no one would be able to determine which were the names and which were the realities" (*Crat.* 432d). Baudrillard already haunts Platonism, which struggles anxiously to refute the virtualities of nominalism and desires to retain a gap (but a gap that links) between names and things.

In a certain sense, it is "within" this infinite gap, this jointure and abyss, that Heidegger, Freud, and Derrida work to transfigure the structure of technologics and Dasein's relationship to it, even though

> This gap between, on the one hand, thought, language, and
> . desire and, on the other hand, knowledge, philosophy, sci-
> ence, and the order of presence is also a gap between gift
> and economy. This gap is not present anywhere; it resembles
> an empty word or a transcendental illusion. But it also gives
> to this structure or to this logic a form analogous to Kant's
> transcendental dialectic, as relation between thinking and
> knowing, the noumenal and the phenomenal. Perhaps this
> analogy will help us and perhaps it has an essential relation
> to the problem of "giving time." (Derrida 1992a., 30)

The possibility of a double creates anxiety, for the double would imply the loss of the possibility of determining difference, especially the difference between truth and its others. This anxiety, which I take to be fundamental to Plato's project, returns at the conclusion of the dialogue, when Socrates—using images of the whirlpool, things that leak like a pot, and a man with a runny nose—summarizes the struggle to relate knowledge and the transitory. Cratylus, becoming not a second Cratylus but a form of Socrates, has the last word: "Very good, Socrates; I hope, however, that you will continue to think about these things yourself" (*Crat.* 440).

Without difference, or with the double (much less the multiply doubled), there is a confusion about demarcations, about what's what, what's real and what is not. This is the enigmatic development that Baudrillard and others have examined under the name of virtual reality, the "simulacra of the simulacra" in which there is no founding "real." At the crossroads of the transepochal, the virtual is produced not just within the order of the *nous* or the *logos*, but is concretized within and through the order of *technē*. Philosophy becomes a cognitive science and Cratylus becomes an android.

When the same is infinitely reproduced—whether by Socrates or in a media-driven society—a kind of terrifying vertigo, a disorientation to the directionality of existence, ensues. As i ek puts it: "The suspension of the function of the (symbolic) Master is the crucial feature of the Real whose contours loom on the horizon of

the cyberspace universe: the moment of implosion when humanity will attain the limit that is impossible to transgress; the moment at which the co-ordinates of our societal life will be
dissolved" (1997, 154). Already we can hear the distant buzzing of
the mechanical bees of capitalism and the droning collapse of the
con- of consignation. The garden of Zapparoni awaits.

There are, however, always treatments for anxiety. As Socrates
responds to his interlocutor,

> I quite agree with you that words should as far as possible
> resemble things; but I fear that this dragging in of resem-
> blance, as Hermogenes says, is a shabby thing, which has to
> be *supplemented by the mechanical aid of convention with a*
> *view to correctness*; for I believe that if we could always, or
> almost always, use likenesses, which are perfectly appropri-
> ate, this would be the most perfect state of language; as the
> opposite is the most imperfect. (*Crat.* 435b; my emphasis)

There is always a prosthetic supplement to be attached to the mimetic
machine to ensure that it functions at what is at least a minimum level
of efficiency. There is always a psychotropic drug, produced by the
pharmaceuticals, ready to allay the anxiety. And likeness, rather than
difference, approaches the state of perfection. Babel will be replaced
with a universal language; communication will be free and clear.

In an extended footnote that accompanies "The Double Session,"
Derrida outlines the logic of Platonic *mimēsis*, especially of the "'internal' duplicity of the *mimeisthai* that Plato wants to cut in two, in order
to separate good *mimēsis* (which reproduces faithfully and truly yet is
already threatened by the simple fact of its duplication) from bad,
which must be contained like madness and (harmful) play" (1981, 187,
n. 14). In this note, which is a kind of synecdoche of the deconstructive
critique of the onto-theological tradition, Derrida lays out a schema of
mimetic doubling and its implications. He writes,

> 1. *Mimēsis* produces a thing's double. If the double is faithful
> and perfectly like, no qualitative difference separates it from
> the model. Three consequences of this: (a) The double—the
> imitator—is nothing, is worth nothing in itself. (b) Since the
> imitator's value comes only from its model, the imitator is
> good when the model is good, and bad when the model is bad.
> In itself it is neutral and transparent. (c) If *mimēsis* is nothing
> and is worth nothing it itself, then it is nothing in value and

being—it is in itself negative. Therefore it is an evil: to imitate
is bad in itself and not just when what is imitated is bad. 2.
Whether like or unlike, the imitator is something, since
mimēsis and likenesses do exist. Therefore this nonbeing does
"exist" in some way (*The Sophist*). Hence: (a) in adding to the
model, the imitator comes as a supplement and ceases to be a
nothing or a nonvalue. (b) In adding to the "existing" model,
the imitator is not the same thing, and even if the resemblance
were absolute, the resemblance is never absolute (*Cratylus*).
And hence never absolutely true. (c) As a supplement that can
take the model's place but never be its equal, the imitator is in
essence inferior even at the moment it replaces the model and
is thus "promoted." The schema (two propositions and six
possible consequences) forms a kind of logical machine; it
programs the prototypes of all the propositions inscribed in
Plato's discourse as well as those of the whole tradition. According to a complex but implacable law, this machine deals
out all the clichés of criticism to come. (Ibid)

Mimēsis is a "logical machine" that, via the copy function, "programs the prototypes of all the propositions . . . of the whole tradition." That, in the shell of a gingernut that Bartleby will keep hidden
away, is the law.

The world is reproduced in thought, and thought, in turn, is reproduced in language. Language, as a necessary medium, produces
the correct image of the double of the world. The machine works via
linkages of likeness (but not identity—although cloning will serve as
a kind of test case) and via a loss of plenitude, an erosion of value,
from original to copy. The copy (the secondary, the derivative) must
simultaneously be recognizable and, in some way, different. Otherwise, there are only reflections of reflections of reflections: a virtual
hall of mirrors in which there are no criteria for distinguishing one
simulacrum from another. There is, finally, no other.

Plato affirms that some types of language behave properly: the
rational language of the *logos*, based on the principle of identity, that
leaves other language games always out of bounds and on the sidelines. But this differentiation within language can occur *only* if good
and proper *mimēsis* is at work, and, as Weber reminds us, "you *cannot* simply separate good *mimēsis* from bad. *Mimēsis* is good and
bad at the same time. That's why it's 'bad,' why it has been considered
with suspicion by moral philosophers from Plato on" (1996, 186).
Rodolphe Gasché expands on this rejection of the dividing difference

that founds philosophy by rejecting all forms of sophistry. "It does not help," he asserts, "to try to . . . differentiate between a good repetition that gives and presents the *eidos*, the ideal and unchanging self-identity of truth, and a bad repetition that repeats repetition, merely repeating itself instead of the living truth. One cannot choose between the living repetition of life and truth and the dead repetition of death and nontruth . . ." (1988, 215).

"Iterability," one of Derrida's names for the necessary condition for any *mimēsis*, implies "*both* identity *and* difference. Iteration in its 'purest' form—and it is always impure—contains in itself the discrepancy of difference that constitutes it as iteration. The iterability of an element divides its own identity a priori . . . it is a differential structure escaping the logic of presence or the (simple and dialectical) opposition of presence and absence . . ." (1988, 53). Explicating this notion, Weber notes that the "*Doppelgänger* is the most direct manifestation of this splitting: the *splitting image*, one could say, of the *self*. The paradoxical twist, however, is that according to the deconstructive graphics of simultaneity, any identity, including the self or the subject, is constituted only and in and through this split, this doubling" (1996, 144).

Where there is a *Doppelgänger*, there will be strange sightings, a summons in the night. There will be deaths and the unburied will walk. And since every entity is self-splitting, always splitting itself open toward the other, which always lodges both within the house and without, everything will obey the logic of the specter. This is neither a new nor an old logic, for it precedes such temporal divisions. Cyberspace, as i ek points out, "merely radicalizes the gap constitutive of the symbolic order: (symbolic) reality always-already was 'virtual'; that is to say: *every access to (social) reality has to be supported by an implicit phantasmic hypertext*" (1997, 143). In fact, his complaint is that "cyberspace is *not spectral enough*. That is to say: the status of what we have called the 'real presence of the Other' is inherently spectral" (155). In the end, working to listen to the phantoms, I will return to the dangers of trying to rid ourselves of the "phantasmic hypertext" in the name of a wish for a certain simplification of the rational, which would like to obliterate the spectrality of the other.

If—to return to the fantasy of a transmission without remainder, without a ghost trailing in its wake—if *mimēsis* assumes the structure of original and copy, then the idea of perfect reproduction is synonymous with cloning, the undifferentiated (re)production of the same in which there is no mark of singularity between the original and the copy. Difference collapses. This is an impossible fantasy, but it is nonetheless one that anxiously circulates when debates about cloning occur, whether in the Platonic text or in postmillennial legislatures. And whenever the logic of original and copy, first and second, is operative, there, too, is a linear logic of time. Obviously, the first comes *before* the second, both ontologically and temporally. First is "higher" and more real; the second is "lower" and less real. The movement of time itself is a kind of erosion of plenitude. One *falls* into time, as from an elsewhere.

This is the logic of metaphysics, whether in a sophisticated or "vulgar" form, which is quite familiar to us all. It is, after all, the apparent form of everyday life: origins originate, time moves from before to after, and causes cause effects. Every entity is itself and not another. The (non)logic of iterability, on the other hand, is unfamiliar, ridiculous, absurd, and irrational. It shatters the law of presence and therefore shatters ordinary notions of temporal progression, cause and effect, and identity. *Mimēsis* and dialectics are always teleological. Iterability is a logic of a (non)simultaneous both/and, an internal division that enables transmission, signification—indeed, any being at all. (And, for Derrida, it is "prior" to Heidegger's "ontological distinction" between Being and beings.)

Iterability names the doubleness (indeed, the plurality) of all possible experience. As the incessantly occurring proliferation of difference within the heart of any eventuality, iterability, as it were, ensures the crossing over of the animate and the inanimate, life and death, when the latter serves as the condition "without which no unit could be exchanged, transmitted, represented, referred to, reproduced, remembered, and so on" (Gasché 1988, 214). And since "the double is the ghostlike manifestation of *iterability* . . . the *splitting image* of the self" (Weber 1996, 144), the dead past encounters us as the living future, and the visage of a ghost will appear.

Ghosts always linger along boundaries that we are frightened to cross. But without this prior address we cannot experience that boundary or attempt—if only by addressing what has addressed us—to speak back. Such a return of the address, such speaking, is difficult, for our skin crawls when we are in the presence of the (un)dead. We are baffled, for what if

> one got to thinking that writing as a *pharmakon* cannot simply be assigned a site within what it situates, cannot be subsumed under concepts whose contours it draws, leaves only its ghost to a logic that can only seek to govern it insofar as logic arises

from it—one would then have to *bend* into strange contortions what could no longer even simply be called logic or discourse. All the more so if what we have just imprudently called a *ghost* can longer be distinguished, with the same assurance, from truth, reality, living flesh, etc. (Derrida 1981, 104)

If so, then language itself, as well all the crossings between so-called dead and living flesh that we are making in the transepochal, will cause us to (s)crawl along the boundaries of the *unheimlich*.

The End of Philosophy and the Beginning of Cybernetics

"End" and "beginning" are, under the sign of the logic of iterability, understood differently. But philosophy, it has been said, is over. Kaput. Broken down, de-composed and analyzed. It has become other disciplines: politics, logic, physics, human resource management, business administration, economics, psychology, cognitive science, cybernetics. Socrates has at last been laid to rest. Quieted at last, just as the Athenians wished. But since philosophy has never been—could never have been and can never be—alive, even though it has been admirably animated, it cannot quite die either. Philosophy, which acted as the general foreman in the construction of the skeletal grid of value that, scraping the sky, separates and thus defines the regions of the animate and the inanimate, the living and the dead, itself is neither.

Or both. Take your pick.

It occurs as the formulation within and of the in-between, and will therefore, of necessity and in order to save itself, speak of the nothing, the abyss, and of bridges. Philosophy always stands on the threshold of the two categories; it operates on a fence, perhaps even *as* a fence, not only marking boundaries but also distributing stolen goods. It is not accidental that philosophy, in one of its disguises—and there is nothing but mask and role at work—develops into hermeneutics, ruled by the god of thieves, among whom there is great honor for a job well done.

Philosophy has a half-life; it exists as a slow vanishing, neither from an original point of plenitude and full knowledge nor from a point of lack that, once science has worked its magic only a little while longer, will one day be happily rounded off and fulfilled. Philosophy is, at every moment of its existence, a vanishing, inscriptions always being read as they are erased and being rewritten by multiple acts of erasure. The ghosts depart and return, as its phenomenal form of being, in the present that twists into the forms of ecstatic temporality. Philosophy itself, represented by that indefatigable questioner, is among those phantoms counted as (un)dead and (non)living. After all, its time is past, over and done with. Where will it go? How will its death continue and be accomplished?

We have other words from the dead who have been precisely recorded and duplicated time and time again. It re-turns. Heidegger once said to a public mirror:

> Heidegger: *Die Philosophie löst sich auf in Einzelwissenschaften; die Psychologie, die Logik, die Politologie.*
>
> Spiegel: *Und wer nimmt den Platz der Philosophie jetzt ein?*
>
> Heidegger: *Die Kybernetik.*
>
> (Heidegger 1988, 102)

And how, here at the end (*Vollendung*) of philosophy, do we get from Plato to cybernetics, command-and-control, surveillance satellites, cloning, artificial intelligence? Perhaps there is nowhere to go. It's all here, waiting. The shades whisper among themselves, waiting for the moment of projective return. The setup is nearly complete. The default for the machine of thought is set at imitation and repetition, but to date it has always produced a flawed copy. It threatens to get out of control.

The aliens are coming, and the apparatus does not yet run on its own. But both technoscience and technocapitalism will labor tirelessly, night and day, to correct this fault, to rid information transfers from any background noise so that all the communiqués can be clear and distinct, copied and stored at will. The stutter in the systematic machinery of (re)production, the snag in the syntax of the fundamental programs: everything can be smoothed out. Technologics, after all, is a logic of mastery that labors (post)industriously to overcome anxiety and make the planetary engines purr peacefully.

Plato, that cagey trickster, compresses the files and teleports the entirety of the mimetic machinery of doubling, the grid of the divided ascent, and rational calculation into the future, where it will be downloaded by, among others, Descartes, Kant, and Hegel, then recoded by Nietzsche and Marx, before it appears in the technical manuals of the bioengineering companies. And once Plato is absorbed into and transformed by Marx's new dialectic, the dream of immortality, as well, must take on a new form. This new formation, one of whose names is "alchemy," will involve, again, the most ancient and the most modern dreams of the modes of production.

II

Ghosts in the Machines

And yet there are moments when the song of machines, the delicate humming of electrical currents, the trembling of turbines in a waterfall, and the rhythmic explosion of motors overcome us with a pride more secret than that known to a victor in his triumph.

—Jünger

4

The Elixir of Life

The hoarder behaves then like an alchemist, speculating
on ghosts, the "elixir of life," the "philosopher's stone."
Speculation is always fascinated, bewitched by the specter.

—Derrida

M any specters haunt the nineteenth century. One of them is
called communism, but there is an uncanny phantasmagoria of other names produced by the encounter between the
ghosts of the past and the ghosts of the future in a present that is
haunted by both. As capital, linked with the sciences, begins to gain
the power to realize ancient dreams of immortality, Nathaniel
Hawthorne constructs "Dr. Heidegger's Experiment" (1837), Mary
Shelley writes not only *Frankenstein* (1818/31) but also the "The
Mortal Immortal" (1833), and, most systematically, Karl Marx, also
employing the language of alchemy, labors to exorcise the ontological reason of Hegel's Absolute Spirit as he elaborates, prophetically,
on the linkage between human and machine beings.

In Hawthorne's story, the old Dr. Heidegger, a "very singular
man," performed a "little experiment" long before the investigations
of Being, Time, and the *Gestell* of modern technology. "If all the stories were true" (1959, 113), the narrator muses, then the old doctor's
study must have been an odd place indeed. "Around the walls stood
several oaken bookcases, the lower shelves of which were filled with
rows of gigantic folios and black-letter quartos." A skeleton was
stored in a closet; a looking glass hung between the bookcases,

where it was "fabled that the spirits of all the doctor's deceased patients dwelt within its verge, and would stare him in the face whenever he looked thitherward" (114). A portrait of his dead fiancée, who had taken one of her lover's prescriptions and died on the night before the wedding, graced the opposite wall.

The study is a place of technics, learning, guilt, desire, and death. In fact, the spirits of the dead appear as mirror images to the gaze of the doctor, staring back at him when he peers into the mirror. His projective reflection calls back the dead from the silvery surface of the mirror of memory that serves as a threshold between the living and the nonliving, who are, nonetheless, able to be summoned and represented. Such crossings between putatively distinct regions of being, now being translated from the speculative—whether of a mythic or philosophical type—to the empiricism of the pragmatic, mark the "period" of the transepochal, which, although it is now accelerating, has been in preparation for millennia.

Dr. Heidegger's chambered text is also a strange studio of words, a place of magic and revivification where the "greatest curiosity" (114) is a book of magic. The narrator, who is never identified, continues to set the scene for "our tale" by describing the doctor himself:

> Now Dr. Heidegger was a very strange old gentleman, whose eccentricity had become the nucleus for a thousand fantastic stories. Some of these fables, to my shame be it spoken, might possibly be traced back to my own veracious self; and if any passages of the present tale should startle the reader's faith, I must be content to bear the stigma of a fiction monger. (114)

Ec-centricity (or perhaps what Derrida calls the "ex-orbitant," a spinning out of orbit that has everything to do with the earth spinning off its tracks) serves to generate fantastic texts, thousands of them, and the narrator is implicated in "shame," describing himself—or herself—as *both* "veracious" and as a "fiction monger." That which narrates always narrates itself as well as its object. And a certain voraciousness is also close at hand. The narrator is, apparently, as exorbitantly eccentric as the strange Dr. Heidegger, and both are the sources of an endless fascination that takes the form of narratives to be told and retold, however unbelievable. Both, it seems, are dealers, merchants even, in fiction. In primitive dreams of the philosopher's stone.

The initial gambit in this story, at first thought to be a "conjuror's trick" (and there is plenty of magic in all of these texts), is to revive a rose that was given to the old Dr. Heidegger by Sylvia Ward, his long dead fiancée. It is cast into a vase brimming with the elixir of life sent by a friend from the magic kingdom itself, Florida. (The story was first published under the title "The Fountain of Youth.") The four guests of the alchemist then drink the aqua vitae and, immediately, there is a "healthful suffusion on their cheeks, instead of the ashen hue that had made them look so corpse-like" (116). The ancient law that sets the course of human life from "ashes to ashes" has been reversed, and the four participants begin to become young for the second time, but experience has added no wisdom. Their old foibles are also revived by the elixir: a focus on coquettish flirtation (by the Widow Wycherly and Colonel Killigrew); the rhetoric of nationalism and patriotism (by Mr. Gascoigne); and the calculating required by economic enterprises (by Mr. Medbourne).

A second chance seems, at least at this first point of intoxication, to be merely a repetition of an earlier template. There is no Socratic advance in wisdom through questioning, but only the monotony of mere repetition, the first time as farce and the second time as farce. Only Dr. Heidegger, living with the knowledge of the tragic, is able to resist the temptation to turn back time. He is, after all, responsible for the death of Sylvia through the misuse of his chymical arts, and only by dying does he have any possibility of making recompense and encountering her on the other side of the mirror.

While Dr. Heidegger sits in his "highbacked, elaborately-carved, oaken arm-chair, with a gray dignity of aspect that might have well befitted that very Father Time, whose power had never been disputed, save by this fortunate company," the others, "almost awed by the expression of his mysterious visage" (117), continue to down the elixir with abandon, until once again the "fresh gloss of the soul, so early lost, and without which the world's successive scenes had been but a gallery of faded pictures, again threw its enchantment over all their prospects" (117). Indeed, they once again *have* prospects other than the grim end of death that has long been, like the many dead that inhabit the mirror of the study, staring them in the face. The "mysterious visage" of time, in its phallic aspect as the law of the father, is cast aside in favor of the pleasures of *enchantment*. The guests prefer the lantern show of youth to the faded portraits on the wall of the study that memory, through the medium of art, has created.

The four revelers first mock the old—their recent selves and Dr. Heidegger—while the latter refuses the lure of youth, remaining the selfsame man in his old age. In a bacchanalian dance, the men clutch at the woman, and "never was there a livelier picture of youthful rivalship, with bewitching beauty for the prize. Yet, by a strange deception . . . the tall mirror is said to have reflected the figures of the three old, gray, withered grandsires, ridiculously contending for the skinny ugliness of a shriveled grandam" (118). The mirror, with its secrets, casts a different angle of light on the *tableau vivant* of those who appear young to their own enchanted vision.

The outcome is predictable. Struggling to win the "girl-widow's" charms, the vase full of the water of life is broken; it spills on a dying butterfly, which then alights on Dr. Heidegger's head. Psyche, always rejuvenated, comes home to the snowy peak of the simulacrum of Father Time, the source of illusion, existence, and death. As the butterfly dies, the rose fades; and Heidegger notes, "I love it as well thus as in its dewy freshness" (119). His seems to be a double love; it is simultaneously of the changing qualities of the rose as well as of the memorialized idea of his own youthful love and guilt, embodied in the flowering rose. The four guests grow old once more, grieving at the changes wrought, and the widow "wishe[s] that the coffin lid were over [her face], since it could be longer beautiful" (119). Although they vow to head immediately to Florida in search of their youth, Dr. Heidegger has his experimental results and turns away from the intoxicating elixir toward the verge of his mirror, as he prepares to step over its edge into the other side of reflection.

Mary Shelley is another nineteenth-century visionary of the transepochal. Like Hawthorne, she writes in order to step beyond the world of banal realism. Her short story "The Mortal Immortal" once again employs the figure of alchemical knowledge in order to tear through the veil of appearances to the other side of life. Winzy, the narrator of the short tale, had been an assistant (a mere 323 years ago) to the alchemist Cornelius Agrippa (1486–1535) and is now attempting to determine whether he is, in fact, immortal. "I will tell my story," he writes, "and my reader shall judge for me. I will tell my story, and so contrive to pass some few hours of a long eternity, become so wearisome to me" (Shelley 1996, 360).

Narration is a way of biding time until the unpredictable, perhaps never attainable, end. Unlike Scheherezade's famous tales, this narration does not keep death at bay, for death in this instance would be a clarifying gift acting to resolve the ambiguity of a *Dasein* structured as mortal-immortal (which prefigures certain forms of the cyborg). It is, instead, simply a substitution for the boredom entailed in an incessant waiting. In this sense it typifies twentieth-century writing organized around names such as Beckett, Blanchot, and Bernhard. (Perhaps those born with surnames beginning with *B* are fated to think the endless deferral entailed by waiting for being.)

All those centuries ago, Winzy, a young man at the time, was in love with a childhood friend, Bertha, who after the death of her parents had been virtually imprisoned by an "old lady of the near castle, rich, childless, and solitary. . . . Henceforth Bertha was clad in silk-inhabited a marble palace—and was looked on as being highly favored by fortune" (361). We are in the midst of a fairy tale. The young man must become an assistant to Agrippa so that he can have a "purse of gold" for his marriage. Gold, presumably, will release the maiden from bondage inside the marble palace. Money will provide liberation. But as Freud has shown, and many others have further explicated, money is shit; shit is a penis; a penis is a baby; a baby is a royal personage. Gold is not the final, liberating term of the series of metaphors (as we will also see with Bartleby), but only another term in the circulation of energy and meanings.[1]

Bertha is beginning to tire of the narrator's inability to "be in two places at once for her sake" (362) and, simultaneously, Cornelius demands the utmost attention from Winzy as he finishes the preparation of a love philter. Grown weary with the watching, the young man "gazed on it with wonder: flashes of admirable beauty, more bright than those which the diamond emits when the sun's rays are on it, glanced from the surface of the liquid . . . the vessel seemed a globe of living radiance, lovely to the eye, and most inviting to the taste." He drinks, believing the elixir will "cure me of love—of torture!" (363). It is the object of perfect fascination, perfect obsession: the fantasy, a radiant globe of wholeness, of a cure from the pains of love. His fate will, at least on the surface of things, be different than that of Dr. Heidegger's four guests, but he will not be able to escape the suffering of the vexed relationship between desire, memory, and time.

Winzy, jubilant with joy, meets Bertha, and desires her more than ever; indeed, he "worships" and "idolizes" her. She asks him to take her away from the life of riches to "poverty and happiness." Luxuries, those monumentalized forms of excrement, are apparently not enough to satisfy her. Thus the girl escapes from her "gilt cage to nature and liberty," not knowing that there are iron cages in the world as well, and that they are not unrelated to g(u)ilt. Bertha and Winzy marry, and he is transported from a "rather serious, or even sad" disposition to a character of "lightness of heart and gaiety" (364). The alchemical pharmacology of psychotropics, long under preparation, seems to work efficaciously.

Cornelius, on his deathbed, calls to Winzy and, once more, shows him the Elixir of Immortality. As if his name were Dr. Heidegger, he commands his assistant to observe the "vanity of human wishes!" (364), reaches for the vessel, and then dies. Winzy begins to wonder whether he is actually immortal or only blessed with an extraordinary longevity, for Cornelius was a "wise philosopher, but had no acquaintance with any spirits but those clad in flesh and blood. His science was simply human; and human science, I soon persuaded myself, could never conquer nature's law so far as to imprison the soul forever within its carnal habitation" (365). Metaphysics, in the sense of that other Dr. Heidegger, is working its work in this passage, but, surely, "nature's laws," the most basic of which is that all that appears in being vanishes as well, are the final unchangeable regulators of the cosmos. The order of death is concomitant with the possibility of life; for chaos to emerge into cosmos, there must be a continuous passing away.

Bertha becomes old and eventually dies, while Winzy, now known as the "scholar bewitched," remains young, restlessly trying to determine his fate. After the passage of centuries, he declares,

> I pause here in my history—I will pursue it no further. A sailor without rudder or compass, tossed on a stormy sea—a traveler lost on a widespread heath, without landmark or star to guide him—such I have been: more lost, more hopeless than either. A nearing ship, a gleam from some far coast, may save them, but I have no beacon except the hope of death. (368)

What, first of all and most of the time, creates a compass for the human passage from nothing to nothing? Death and its other; death as a portal to its other.

Death, for mortals, grants a contoured horizon to existence and thereby acts as an existential compass for human activity, enabling a certain type of evaluation. In the light of the certainty of impending death, what shall I do? How shall I spend my limited time? But the nineteenth century is a period of enormous transition, and in the mortal/immortal's plant there is already the echo of the futural voice of Nietzsche's madman, who carries a lantern through the funereal churches. "What did we do when we loosened this earth from its sun? Where does it now move? Where do we move? Away from all suns? Don't we stumble on without any goals? Backwards, sideways, forwards, in all directions? Is there still an above and below?" (Nietzsche 1971, 168). The death of God collapses all directionality of ethics and meaning for the lantern bearer; and, so too, without the assurance of his own being-toward-death, Winzy roams without direction, without any particular intention except to find out whether he is, after all, a mortal.

This directionlessness brings him—and all of Shelley's work—into the closest proximity to our own period in which the directionality of history, for so long provided by the underlying Judeo-Christian myth of a progressive time augmented by the homogenous space-time of classical physics, has been torn from its tracks. As that other Dr. Heidegger writes in his analysis of the broken lantern: "The pronouncement 'God is dead' means: The suprasensory world is without effective power. It bestows no life. Metaphysics, for Nietzsche western philosophy understood as Platonism, is at an end. . . . [N]othing more remains to which man can cling and by which he can orient himself" (1977, 61). But human beings do not do well without orientations, and in the transepochal it is cybernetics in the broadest sense that, as both the substitute for and the fulfillment of metaphysics, comes to reign. But although the *kybernetes* is a steersman, the rudder is either fixed in one direction—that of a capitalist technoscience that governs society—or it has been dislodged from the vessel of the world.

The death of God is conjoined with the fantasized death of human death—the most ancient fantasy—and in this deathly deathlessness *Dasein* becomes fundamentally disoriented. This disorientation is tantamount to being oriented by nothing, nothing at all, and thus there emerges, as if from a grave, nihilism, that "most uncanny of all guests." In the context we are exploring, both Hawthorne and Shelley must host the uncanny guest that has arrived unbidden at their doorstep (perhaps in need of lodging, perhaps to revenge an old wrong—who knows?) and they must accept their responsibilities as those cultural fabulists called *writers*.

They are neither Renaissance alchemists nor philosophers of the truth. They are responsible neither to the retort, to the reason of metaphysics, nor to science. Rather, they are fictionalists, and fiction—which is neither true nor false and is both true and false—is one of the primary sites of the presencing of the uncanny.[2] In this time of transit when time is being fundamentally questioned, we have become, like the earth itself, peripatetics. The fictive expands, especially in the twentieth century, to become, through the transformers of the information sciences, the space of virtuality. Modernity bleeds into postmodernity, more haunted than ever by the uncanniness of the (k)nots between life and death. It is but a blink of the eye, or less, from alchemical narratives to the cyborg.

As Winzy continues to reflect upon the dilemma of his end(lessness), he recalls that he drank "only half" of the beaker of the elixir. But can, he wonders, immortality be divided? Can time be fractioned? "I often try to imagine," he muses, "by what rule the infinite may be divided" (Shelley 1996, 368). What is half of the infinite, half of immortality? The whole of each. There is no "rule" that governs the division of the infinite; and, only the finite world of phenomenal appearance can be cut and dried, measured along a line of the large and the small, the heavy and the light. There is that which gives measure to the earth, the ratio to all things, but that itself cannot be measured by any ratio. It is the (k)notting, the jointure. Winzy considers suicide, but there is of course the question of whether this is even a possibility. *Can* an immortal take his or her own life? Much of the story toward its end assumes that Winzy can, if put in the path of destruction, lose his life, but that is not immortality. He speaks of

> an expedition, which mortal frame can never survive, even [one] endued with the youth and strength that inhabits mine. . . . Before I go, a miserable vanity has caused me to pen these pages. I would not die, and leave no name behind. . . . or if I survive, my name shall be recorded as one of the most famous among the sons of men, and, my task achieved, I shall adopt more resolute means, and, by scattering and annihilating the atoms that compose my frame, set at liberty the life imprisoned within, and so cruelly prevented from soaring from this dim earth to a sphere more congenial to its immortal essence. (369)[3]

Writing, in this instance, is also a space for the scatalogically narcissistic fantasy of omnipotence: "[L]et me leave my remains, my inky or phosphorescent droppings, here on the screen of this page and I will be remembered forever." Let me leave my name behind, so that it can always (re)appear on a fresh page, as a reflection on a fresh reader's eye. Let me gather and transmute the syllables of Wollstonecraft-Shelley and become, if only for a moment in which writing-reading lasts as long as eternity, Winzy.

He promises to destroy the "frame" and liberate the true life of the spirit. Western metaphysics presses relentlessly down on Winzy's fevered mind and reproduces the body-spirit division in its most typical form. And yet the question remains: what of Winzy is immortal? Surely not just the spirit, with its ungainly name of Winzy? That, surely, would be whimsical. The possibility of immortality creates a concomitant and powerful will-toward-death, an "abhorrence" for life, something we should take note of in this period of genetic engineering, cloning, the emergence of artificial intelligence, and all the phantasmagoria of contemporary weaponry, chemical and otherwise.

The elixir of life cocreates the elixir of death; the conditions for the possibility of immortality cocreate the conditions for the possibility of absolute annihilation.

Thus, the story of Winzy's fantastic, chemically induced encounter with the infinite ends, in an ascent—although the descent, a revaluation of values, has already been set in place by the romantics—toward a realm of the "immortal essence." Winzy, his voice fading into the blankness of the page, is of course only a virtual avatar, a ghost that can be called into being only by an incantation that is a short story. Magic has been itself transmogrified into writing and into technology. All these hoarders, haunted by the prospect of immortality, have attempted to bend science to their wills. Attempted to hoard their morsel of life. But though the story of the bewitched scholar is finished, there is yet the little morsel, the small leftover, of the date: 1833. Time is still marked; immortality has not yet set in for the writer or her readers, for time can be, still, fractioned, calculated, and posted on the page.

5

The Immortality Machine of Capitalism

Marx does not like ghosts any more than his adversaries
do. He does not want to believe in them. But he thinks of
nothing else.

—Derrida

In 1833, Karl Marx turned twelve years old and alchemy, Frankenstein, and capitalism awaited his coming. What Marx would come to understand better than anyone else in the nineteenth century was the historical *sociality*—the "social brain" of emergent capitalism—of all human events, and it was Marx who first demonstrated how the alchemical dream of immortality had been displaced from the individual study of Agrippa or Dr. Heidegger to the entirety of the social world itself. Capitalist society itself had become an enormous alchemical laboratory, working incessantly to transform the dross of unused nature into the gold of surplus value. But elixirs, death, vampires, and ghosts—all remnants and revenants of the so-called past—could not be kept out of this laboratory any more than from those of earlier social formations. And they would be coupled with the machine.

In the *Eighteenth Brumaire of Louis Bonaparte* (1851), Marx notes that "Men make their own history, but they do not make it just as they please; they do not make it under circumstances chosen by themselves, but under circumstances directly found, given and transmitted from the past. The tradition of all the dead generations weighs like a nightmare on the brain of the living." Just as people are beginning to liberate themselves, they take on a "time-honored disguise and [a] borrowed language" in order to present the "new scene of world history." This new scene, Marx says, is a "conjuring up of the dead" (1978b, 595).

Theater, séances, philosophy, political economy, dreams, and languages are all conflated in the attempt to express the radically new and its relation to the ancient as time pivots into a different historical era, even as its participants are "set back into a dead epoch" (596). The present takes on the form of the past; the past strides into the present as a "caricature." And the "bourgeois order, which at the beginning of the century set the state to stand guard over the newly arisen small holding and manured it with laurels, has become *a vampire* that sucks out its blood and marrow and throws them into the *alchemistic cauldron of capital*" (611; my emphasis).

Vampires roam the earth; capitalism is a cauldron of alchemical ferment. This is a witches' brew indeed, and the philosopher's stone must lie close at hand, if still hidden, waiting until the right moment to make its appearance. Norman O. Brown, in *Life against Death*, explored this displacement in Marx, reminding us that

> With the transformation of the worthless into the priceless and the inedible into food, man acquires a soul; he becomes the animal which does not live by bread alone, the animal which sublimates. Hence gold is the quintessential symbol of the human endeavor to sublimate—"the world's soul" (Jonson). The sublimation of base matter into gold is the folly of alchemy and the folly of alchemy's pseudosecular heir, modern capitalism. (1959, 258)

He then cites several of Marx's references to alchemy in *Capital* and concludes that "Freud's critique of sublimation foreshadows the end of this flight of human fancy, the end of the alchemical delusion, the discovery of what things really are worth, and the return of the priceless to the worthless" (259).

But sublimation, like the go(l)d standard, is in essence the process of transmutation. It is the incessant movement of signification, regardless of how much the ego consciously desires to translate permanently from "low" to "high," from the flux of desire to the stability of an established *logos*, meaning, or value. The sublimated will always de-sublimate; the inflated value will always deflate. Freud's critique *does* illuminate the interior workings of the imagination—including the deflective sublimations practiced by capitalism—but the "alchemical delusion," since it is a necessary fiction, will continue to be reproduced in different forms.[1] The end of ideology, dream, or myth, is the beginning of a new, or newly framed, ideology, dream, or myth. What Baudrillard calls the "vital illusion" is requisite to all forms of the production of meaning, which may be precisely that which is under threat in the transepochal.

Alchemy has always required a workplace and instruments, and it is no different in the space of modern capitalism, its "pseudosecular" heir. The workplace, with laptops and cell phones, can now be anywhere, and the tools of capitalism, machinery in its many forms, are inseparable from the dream of immortality. It is in the *Grundrisse* (1857–58) and its heir, *Capital*, that Marx most presciently lays out the complex development between human beings and machines that will govern the sociality of work from the nineteenth century to our own period of PCs, the development of IBM chessmasters, and the literalization of virtualized reality. As he already saw in 1857:

> Once adopted into the production process of capital, the means of labor passes through different metamorphoses, whose culmination is the *machine*, or rather, an *automatic system of machinery* (system of machinery: the *automatic* one is merely its most complete, most adequate form, and alone transforms machinery into a system), set in motion by an automaton, a moving power that moves itself; this automaton consisting of numerous mechanical and intellectual organs, so that the workers themselves are cast (*bestimmt*) merely as its conscious linkages (*intellektuellen Organen*). (1973, 279)

Labor becomes determined as the raw material, the "lead," which capital transmutes into something of greater value for the system as a whole. The traditional concept of the soul, "a moving power that moves itself," and Hegel's Absolute Spirit unfolding itself in history have been displaced and replaced by Marx with the autonomous, self-governing system of production in which the worker—now tantamount to any human being—becomes merely an appendage of the machinic system.

The network requires, early in its evolution, consciousness as a form of regulating the connecting nodes of the network of machines. Eventually, however, such consciousness—or what Marx also calls a "living accessory" to the system of dead capital—will also become superfluous. With the advent of artificially produced intelligence that can re-create versions of itself and project itself into its own future, human consciousness will be remaindered and cast aside as junk. We can imagine that human history will become the junkyard of the autonomic system of thinking machines; perhaps it will be useful now and again for parts, an idea or a current of emotion, an archive of the most primitive beginnings. Perhaps humanity will become a kind of museum of natural history that continues its development as a subspecies; or, perhaps, preautomaton history will serve merely as a salvage yard of memories that lie forgotten after being wrecked, rusting in the sun and rain of the millennia.

This is not, of course, Marx's conscious political or economic agenda as the machine age prepares for the liberation of the classless society. Nonetheless, perhaps Marx was in fact the prophet of the end of humanity in ways that he never envisaged. The system of the automaton belongs not to the individual—machines do not make the individual producer's life better—but to organized capital. It is the machine and the "raw material" that are brought into a fundamental relationship, with the worker only being the (temporary) connecting link. The machine of capitalism is *not* a tool wielded by a worker on an object of work. Instead, it is "posited in such a way that it merely transmits the machine's work, the machine's action, on to the raw material—supervises it and guards against interruptions (*überwacht und sie vor Störungen bewahrt*)" (279).

"Supervision," of course, now resonates through the many analyses of the "surveillance society," but Marx is already suggesting that it is part of the essence of capitalism to have such a function built into its own machinery; the panopticon is built into the system from the beginning. Human beings are determined as the method of the specular reflection of capitalism's systemic transmutations. Capitalism gathers the machines and the humans together into, explicitly, a machine-human system. We are commanded to keep an eye out for it, on it, by means of it. Watch out: that's the imperative.

This "watching" is at the core of western metaphysics, and the technology of surveillance is enabled by the philosophemes that serve as the context for the production of satellites, mobile phones, and all the rest. In a translator's note to Heidegger's "The Turning," Lovitt remarks that *"Wahren*, ordinarily understood as to watch over, to keep safe, to preserve—and with it *Wahrnis*—clearly carries, simultaneously, connotations of freeing, i.e., of allowing to be manifest. The same connotations are resident in all the words built on *wahr*. They should be heard in *Wahrheit* (truth) . . ." (Heidegger 1977b, 43). And, in a passage from "Science and Reflection," Heidegger himself explains that "In *theoria* [the Greek term] transformed into *contemplatio* [the Latin] there comes to the fore the impulse, already prepared in Greek thinking, of a looking-at that sunders and compartmentalizes. A type of encroaching advance by successive interrelated steps toward that which is to be grasped by the eye makes itself normative in knowing" (1977b, 166). The eye, in other words, produces its own discourse on method within metaphysics, but the field of vision cannot be completely mastered. There are, at least for the time being, blind spots.

Marx urges us to be especially careful of the *interruptions, the disturbances*, for these will indicate fault-lines and failures within the network of the system: "[E]very interruption of the production process acts as a direct reduction of capital itself" (1973, 283). Capitalism's *time*, founded on the process of communicative productivity, desires itself to be seamless and smooth, unbroken.[2] The machines should *purr* with their own productive power. The whole system should run on its own—which is also Freud's early fantasy in the *Project* about the psychic apparatus of the mind—and without interruption, for any interruption, any unplanned breaks, decrease the value of the overall capital. Subjects will have the immediate gratification of consumption, and objects will be made ready for a "just in time" inventory only within a certain interpretation of time and being. As the second Dr. Heidegger notes, "Machine technology remains up to now the most visible outgrowth of the essence of modern technology, which is identical with the essence of modern metaphysics" (1977, 116). Marx, in these few pages from the *Grundrisse*, is delineating the metaphysics of an entire epoch.

In this epoch the machine becomes the "animator" and the "virtuoso, with a soul of its own (*eine eigne Seele*) in the mechanical laws acting through it, consuming coal, oil . . . to keep up its perpetual motion" (1973, 279). Modernity ransacks the past and all the earth's resources. The continuous mobilities of a fast and efficient productivity are not recent additions of the information age, but have been part (however nascent) of capitalism since its inception. And there is the usual alchemical template at work as the machinery of capitalist society takes in the base materials, transmutes them into something of greater value, and then discards the waste (factories, outdated systems of production or surveillance, ideas, and employees). Everything that can be sold as remainders is sold, and the rest, the unremaindered remainders, the leftovers, are either destroyed or dumped in the landfill. (Such places are represented by William Gibson's junkyard settings of the Factory and Dog Solitude, among many other postapocalyptic, which is to say *present*, sites in the contemporary arts.)

Marx continues the description of the early industrial worker: "The science which compels the inanimate limbs (*unbelebten Glieder*) of the machinery, by their construction to act purposefully, as an automaton, does not exist in the worker's consciousness, but rather acts upon him through the machine as an alien power, as the power of the machine itself" (279). Science and capital, like Dr. Frankenstein, animate the dead machinery of life. Like a puppeteer, it moves the inert limbs of the machine system that then acts "purposefully," like an automaton. Artificial intelligence, in other words, is not primarily a *product* of the capitalist social world—though it is that too—but, rather, is *constitutive* of that world.

Capitalism is structuring of the earth as a form of AI, complete with neural pathways, self-reflexivity, reproduction, and the move from carbon to silicon. Moreover, within that system of active intelligence, human beings—the "worker" is, in the twenty-first century, not an economic class as much as a species division—provide (at least for the time being) the energy necessary to run the machine, and, as such, are acted upon as by an "alien power." The machine is an alien because it is alienated from the production of the worker; it comes as if from "outside" the field of the worker's competency and theoretical understanding. And if the machine is alien, then the worker is an alien as well: split and segmented, compartmentalized into multiple niches that do not necessarily have any meaningful relationship with one another. The machine overwhelms the insignificant actions of the individual. The automaton downsizes the worker, or, as Marx has it, "The increase of the productive force of labor and the greatest possible negation of necessary labor is the necessary tendency of capital. The transformation of the means of labor into machinery is the realization of this tendency" (280).

The "social brain"—all the accumulated knowledge and power of the collectivity of society—is absorbed not into labor, but into circulating capital (for example, electronically driven transnational markets). The fixed form, provisionally bound into place by the network of the social means of production, is broken apart to circulate in other "equivalent incarnations" (280), although no incarnation is quite equivalent to its analogues. The dialectic of the automaton, always a machine but one that in the past required human labor, keeps indefatigably at its work of consolidating, destroying, and reconsolidating. As an autonomous machine, it realizes itself in a new medium in which it may be able, at last, to forgo the "living accessories" called human beings, to do without the prosthesis of *Dasein*. When the machines write the history of their own coming-into-being, there will be a Nietzschedroid who proclaims that "the creator is dead and we have killed him," but there will not be the ensuing sense of guilt, endangerment, and the nihilism that bedeviled the nineteeth and the twentieth centuries.

The bottom line of the capitalist social brain demands, but this is not yet at the end of the line, that the worker be jettisoned from the grid of accountability that structures the possibility of a bottom line and be cast toward the abyss, not only of unemployment, but of nonbeing and death. The libidinal economy in which value(s) circulate founders on the powers of Thanatos that dissolve the individual back into his or her inorganic, component parts. The living regresses to the dead; the animate to the inanimate. And, in the other direction, that which had been assumed to be dead materiality turns itself on and devours the living limbs in order to rejuvenate itself and grow stronger. It is as if the worker as individual producer stands at a crossroads between the animate and the inanimate, and is momentarily transfixed at that crossing by the needs of capital to produce and reproduce.

The system of the machine occurs, for Marx, only when industry has grown large enough to magnetize resources toward itself and when "all the sciences have been pressed into the service of capital," when "invention becomes a business," not the cliché of the eccentric in his workshop but the result of research groups organized by industry's needs and desires: by profit. Marx, however, focuses on another aspect of the machine-human relationship, claiming that the primary "road" of its development is "dissection (*Analyse*)—through the division of labor, which gradually transforms the workers' operations into more and more mechanical ones, so that at a certain point a mechanism can step into their places" (283). Technocapitalism analyzes apparent wholes into smaller and smaller units, and these units are all mechanizable. In the twenty-first century, this division of labor has of course far exceeded the factory model, and, through various forms of the technologies of the microscopic, undone the body, the voice, the personality, and cognition.

As Ernst Jünger observed, the technologized mind that works to "negate the image of the free and intact man . . . wanted units to be equal and divisible, and for that purpose man had to be destroyed" (1960, 141). The artifice of technique is not an add-on or a plug-in; it is always already within the animated whole, and it always everywhere is constitutively breaching that ideal. With the emergence of the social brain and the automaton of capitalism, however, the artifice is exteriorized as a more encompassing exoskeleton and is literalized as electronic intelligence. "What was the living worker's activity," Marx observes, "becomes the activity of the machine. Thus the appropriation of labor by capital confronts the worker in a coarsely sensuous form; capital absorbs labor into itself, 'as though its body were by love possessed'" (1973, 283).[3] In other words, the worker gets fucked. In the affair between individual and system, the system seductively lures the worker into its webworks, but the lure is poison. Eros, within the vampiric economy, leads only to the convulsions of death.

The animate, because it contains the constituent elements of the nonliving within itself, is replaced by the inanimate; but, in its turn, the inanimate begins to take on the very characteristics of animation. There is a crossover in which the opposition between the two, the very opposition that has traditionally grounded the distinction between the living and the dead, the organic and the inorganic, splits asunder, then recombines in a new form. The cyclotron of capitalism and science takes command, splitting all oppositions into elemental units. There are, however, ultimately no "elemental units"; there is only the field in which so-called units explode and reconstitute in virtual clouds of cultural formations that still, occasionally, seem quasi-permanent, but that are in fact always tearing-gathering.

But what, exactly, is the material that is being torn as it gathers, that gathers as it is rent? It is difficult to say. When, for Marx, the "living time of labor" is exchanged for the objective labor of machines, all symmetry between labor time and the development of capital is destroyed. Capital is no longer *measured* by the workers' time on the job, but by the productive abilities of machines and the force of the agencies that set those machines in motion, agencies whose "'powerful effectiveness' is itself in turn out of all proportion to the direct labor time spent on their production, but depends rather on the general state of science and on the progress of technology" (284).

"Labor," however, may no longer be the fitting word, for "[l]abor no longer appears so much to be included within the production process; rather, the human being comes to relate more as *watchman* and *regulator* to the production process itself" (284; emphasis added). Eventually, nothing is made by the human hand except machines that will make what is needed. All that is needed is the "living accessory," the "conscious link" in order to write the code, flick the switches, monitor the gauges, and interpretively redirect the output. Humans are pressed outside of the system of direct production and become regulators of one sort or another. The worker, Marx says, "steps to the side," and in this "transformation, it is neither the direct human labor he himself performs, nor the time during which he works, but rather the appropriation of his own general productive power, his understanding of nature and his mastery over it by virtue of his presence as a social body—it is, in a word, the development of the social individual which appears as the great foundation-stone of production and wealth" (284).

The individual—in particular, the postenlightenment individual of democracies and capitalism—is transformed into a "social individual" or a "social body." And it is that unit—as company, research team, university, government—which becomes the unit of economic production. The single individual "steps aside," not out of some alleged "free choice," but because he or she is forced to the margins, and beyond, of production. This single individual, for the most part and most of the time, to echo Heidegger's characterization of the "They" that we are, does *not* have productive power or an understanding of nature, much less "mastery" over it. Only the collective, which forms and becomes a database—a base for the circulation of information for the sake of moneymaking—has such powers. The single individual is valuable, able to have his or her value measured, in so far as he or she can contribute to the system of production through some capacity to transform nature, not, any longer, directly, but indirectly through the symbolic and regulative functions of science, economics, or the contributions to the knowledge economy.[4]

When the autonomous machine-system becomes necessary for the forces of production, *time changes*. No longer, Marx argues, is labor time the requisite necessity for the creation of wealth, but—since it takes far less human labor time, once capital is fixed in machines, for production to occur—time is "set free," and individuals, if at first only a few, are set free to develop their individuality in any way they see fit. In a paradoxical twist within the Marxist utopia, the individual is obliterated as laborer in order to be re-created as a self-actualizing individual or as a spectator, an omnipresent fan of entertainment.

"Capital itself," Marx continues, "is the moving contradiction, [in] that it presses to reduce labor time to a minimum, while it posits labor time, on the other side, as sole measure and source of wealth. Hence it diminishes labor time in the necessary form so as to increase it in the superfluous form; hence posits the superfluous in growing measure as a condition—question of life or death—for the necessary" (285). The dialectic of contradiction (and what a strange logic this is) produces the "superfluous" as a condition for the "necessary" and this is a "question of life or death." No wonder these are the material conditions that are ready to "blow this foundation sky-high" (285). Capitalism, incorporating the violence of boom and bust, volatilizes every relationship.

The superfluous has bled into the necessary; the necessary into the superfluous. The animate has become inanimate as the inanimate becomes animated. As Marx notes: "Nature builds no machines, no locomotives, railways, electric telegraphs, self-acting mules, etc. These are products of human industry, or of human participation in nature. They are *organs of the human brain, created by the human hand*; the power of knowledge, objectified" (285). The brain and the hand become masters, in a sense, of that which gave rise to them, nature. But now this form of nature is "objectified," separated and established as an independent entity from the subject that at first is necessary for its being. As the second Heidegger paraphrases this observation, everything—including the human thing—is transfixed as a "standing reserve" that has been put "on call."

The institution of the automaton, in which profits can be made all day and everyday, without even the necessity of a human watchman, changes the nature of the human experience of time as it changes the meaning of wealth. Instead of "labor time" being the measure of value, it becomes "disposable time." The time of work, it appears at first blush, can be done away with; we can dis-pose of and unposition time as we see fit, for the good of the individual. But, once again, this is simply a mirage, for as soon as there is disposable time, supposedly a time for freedom, Marx declares that it becomes surplus labor time and there is a "positing of an individual's entire time as labor time, and his degradation therefore to mere worker. . . . *The most developed machinery thus forces the worker to work longer than the savage does, or than he himself did with the simplest, crudest tools*" (287). Our computers, like Bartleby, run a day and night line.

Paradoxically, we are freed up by our machinery to either become unemployed or work longer hours. Capitalism always ups the ante in the casino of the global economy, and what is always at stake is time. *Time is money*. The individual, no longer the direct basis of production or wealth, becomes determined as the *"suspended individual, i.e., as social labor (als aufgehobne einzelne, d.h. als gesellschäftliche Arbeit)"* (288). The individual, both a necessary and a superfluous category in the dialectic of technocapitalism, is *suspended* along the wirework of the system, on call to be used in the workforce, to be cut free to fall, or to dissolve in the alchemical solution brewed by the social mind.

For Marx, both the system and the subject within the system have a destiny: their own vanishing. Capitalism will disappear when the "last form of servitude assumed by human activity, that of wage labor on one side, capital on the other, is thereby *cast off like a skin*" (291). The dialectical-historical mechanisms that constitute capitalism produce machines that mechanize production "beyond a certain point" and beyond that point production, burdened by falling profits, turns self-destructively on itself. A "new man" is being forged in the machinery of capitalism who will, like a snake molting, cast off the old and put on the new. This is not just a metaphor, for part of the strategy of technocapitalism is that the old body will learn to build new parts for itself, reconstructing itself as it decays.

This moment will have the characteristics of an apocalypse, a time in which what has been hidden will emerge, after long preparation in the economic-technical alembic of capitalism, into the light of historical appearance.

> These contradictions lead to explosions, cataclysms, crises, in which by momentous suspension of labor and annihilation of a great portion of capital the latter is violently reduced to the point where it can go on. These contradictions, of course, lead to explosions, crises, in which the momentary suspension of all labor and annihilation of a great part of the capital violently lead it back to the point where it is enabled [to go on] fully employing its productive powers without committing suicide. Yet, these regularly recurring catastrophes lead to their repetition on a higher scale, and finally to its violent overthrow. (292)[5]

History works by means of dismantling old forms of socioeconomic organization and leads onward to a "new foundation," although all "foundations" except the eschatalogical last must be, given the nature of the historical dialectic, provisional and therefore not essentially foundational. Marx does not explicate the idea, but he seems to point past the stage in which machines—and machines are the synecdoche for capitalism and objective labor as a whole—are alienated from the workers and toward an economy in which the changed foundation of economics would rearrange the machine-human relationship. The alienation would be overcome, and the hostility "contained" within the "otherness" of the machines would be resolved.

Through cybernetics, history has instituted this next stage by combining forms of the human and nonhuman into a provisional "new foundation." This new form of existence apparently overcomes the hostility inherent in alienation, at least in its simple "classical" form, by erasing the line that separates and distinguishes human from nonhuman (both animal and inorganic forms). The line can be crossed in either direction: humans can incorporate animals and machines (baboon hearts and pacemakers), and animals and machines can begin to incorporate human characteristics (ears, genes, language, thinking). The end of human history, and the beginning of the posthuman, would necessarily entail a humanization of the machine and the mechanization of the human.

The Spectrality of the Nineteenth Century

Phantoms haunt the nineteeth century, memories of suffering and injustice, of a longing for death as well as a longing to be done with death, and these phantoms are projected forward into the twenty-first century. Capitalism, which comes into its "own," is inseparable not only from rational quantification and the development of industry, but also from alchemy, vampires, ghosts, aliens, and automatons, and from the dead animating the living (which confounds the logic of all such terms). The entire baroque bestiary of the irrational courses along every pathway of the enterprise that encircles, and breaches the circle, of the globe, but there is one more term to be added to this series: organism. The unity of the system, Marx argues, exists not in the living workers, but in "the living (active) machinery (*in der lebendigen (aktiven) Maschinerie*), which confronts his individual, insignificant doings as a mighty organism (*als gewaltiger Organismus*)" (Marx 1973, 279). The living/dead machine is (not) biological.

The monster has come to life, devouring the workers, and it acts like an organism. It is within the vastness of this *like*, which opens up the space of metaphorical substitution, that the technologics of the transepochal operate. This shift to the language of the organism is initially jarring, but Marx is not being incautious with his language. After all, if the machine system produces and reproduces itself, acts with purpose, is animated by knowledge (science), and is far more virtuosic than its human minions, then it is characterized by attributes that we usually use to describe intelligent organisms. In 1857, Marx already envisioned a crossover of disciplines, knowledges, and competencies that would produce the fields of biotechnology, in which *bios* and *technē* interweave, cross-fertilize, and breed new creatures. The biosphere will be radically technologized, and the technological, as the inanimate and artificial, will become biologized. And this occurs as the world is being commodified.

Derrida, analyzing the table in *Capital* that stands as Marx's example of the mystification that comes along with commodification, summarizes these multiple crossings:

> Facing up to the others, before the others, its fellows, here then is the apparition of a strange creature: at the same time Life, Thing, Beast, Object, Commodity, Automaton—in a word, specter. This Thing, which is no longer altogether a thing, here it goes and unfolds, it unfolds itself, it develops what it engenders through a quasi-spontaneous generation (parthenogenesis and indeterminate sexuality: the animal Thing, the animated-inanimated Thing, the dead-living Thing is a Father-Mother). . . . (1994, 152)

This ghostly logic, this hauntology, governs Marx's discourse as it does the entirety of metaphysics, but the emergence and domination of capitalism speeds up the alchemical brewing process that produces the volatile, flammable spirits that rush so potently to our heads.

We exist *as* the explosive splintering of the old table(t)s. The traditional divisions will have been suspended and this is the transepochal moment of waiting, without knowing what we wait for, in which we find and lose ourselves, our histories. The air is very still. This *suspension* marks the in-between moment of waiting, of hoping, of preparing for what is to come. Martin Nicolaus, the translator of the *Grundrisse*, has commented about the use of "suspension" in Marx, and his words are worth citing in full:

> At a certain point [in the dialectic of the "free market"] occurs that which Hegel and Marx call the *Umschlug*—the abrupt,

leap-like inversion or overthrow, in which the previous barrier, the identity, law of equivalence etc. is negated, the underlying contradiction is *suspended*, and the whole is transformed into its opposite, with identities and contradictions of a different order and on a higher level. A word about *suspension*. It translates—Marx himself uses it to translate—Hegel and Marx's term *Aufhebung*. Hegel took delight in the word, as it expresses in ordinary language precisely two opposite senses at once: "it means as much as to preserve, to sustain and at the same time as much as to let cease, to make an end." The English "suspend" has precisely the same contradictory senses; as for instance in commerce "it means to stop (payment) while in music the sense is to continue, sustain (a note), and in bureaucratic administration (as in school systems) it means both at once. Hegel was particularly at pains to point out the difference between suspension and annihilation; that which is suspended has not become nothing, but continues on as "a result, which has come out of a being; hence it still has in itself the determinateness out of which it comes. . . ." (Marx 1973, 32)

Dasein is being-suspended in a universal solvent, a solution called technologics, metaphysics, capitalism, the globalization of planetary cultures. Since, however, it is the solvent in which everything is being-dissolved, it cannot, finally, have a name; or, at the very least, every name must be provisional in the extreme.

Perhaps we can only call it a *flare*: for help in an emergency, to see the ground in front of us, as a blind signal that we are here. The crossover of mortals; the presencing of thing-animal-ghost-organism; the artificial becoming natural and the natural becoming the artificial. An intimate confluence of the most ancient and the most new, as they mix and mingle to create a different spirit-body that will, wandering in need of an other language, haunt the coming century. A flare lights the night sky.

6

Bartleby the Incalculable

The shadow is a manifest, though impenetrable, testimony to the concealed emitting of light. In keeping with this concept of shadow, we experience the incalculable as that which, withdrawn from representation, is nevertheless manifest in whatever is, pointing to Being, which remains concealed.

—Heidegger

The Cistophorus as a Call to Writing

Plato responded to the call of the distinguishability of numbers and letters, constructing a matrix of value that was, after its incarnation as the Hegelian *Weltgeist*, overturned by Marx, who responded to the call of labor, alchemy, money, justice, and a table that was not a table. Already for Marx, the world was haunted by the frozen desire of commodities, fetishes, the vampire of capital, and specters that roamed Europe, specters possessing the workers of the world with a call for unity and revolution. Understanding that the machinic and the human were to be symbiotically linked through the mechanisms of organicity, he wandered the world, fleeing the law. He sought a home, and, almost—not quite, but almost—ended up as a newspaper writer in New York City, that site of the utmost calculability which is, for this very reason, haunted by the incalculable. This, too, will summon thought.

In a discussion of how thinking inclines toward the event of withdrawing, Heidegger argues that Socrates, by placing himself into the draft of that withdrawing, was the "purest thinker of the West," and this is why "he wrote nothing. For anyone who begins to write out of thoughtfulness must inevitably be like those people who run to seek refuge from any draft too strong for them." Thus, after Socrates, all philosophers (indeed, all writers) are "fugitives," for "thinking has entered into literature" (1968, 17–18). Elsewhere in the same text, he claims that "Literature is what has literally been written down and *copied*, with the intent that it be available to a reading public" (134; my emphasis).

In a certain sense, Bartleby, too, "writes nothing," refusing to copy and therefore not participating in the literary that relies on mimetic machinery and a public. And yet he, like Socrates, has a scriptorial other who is willing to do the necessary work, whether it is called "philosophy" or "literature." This thin sheaf of fiction, really only the merest ghost of a story, concerns itself with drafts, money, rationality, machines, the nothing, counterfeiting, and a watery crypt that is a reservoir for thinking. Technologics is already in full gear.

Herman Melville's *Bartleby the Scrivener: A Story of Wall Street*, published in 1853, begins as a wisp in the wind for the narrator, who in order to tell the tale must become a ghost-writer:

> I believe that no materials exist for a full and satisfactory biography of this man. It is an irreparable loss to literature. Bartleby was one of those beings of whom nothing is ascertainable except from the original sources, and, in his case, those are very small. What my own astonished eyes saw of Bartleby, that is all I know of him, except, indeed, one vague report, which will appear in the sequel. (Melville 1990, 3)

Bartleby is without explicable origin. The narrator's writing, whatever it turns out to be, cannot match the "loss to literature." Not a biographer or a man of letters, the narrator has no origin but what he has seen with his own "astonished" eyes, and, as he says before he has well begun, the report that will appear in the "sequel." In the beginning is the afterthought, the afterword. The sequel is included in the origin, and, as usual, the logic of the supplement, while necessary to the operation of the narrator's rendition of Bartleby, will not shed any definitive light on the enigma. It will not, finally, lay any ghosts to rest, but, instead, wake more of the restless shades from the grave pages of capital, justice, and the pyramidical tombs.

Having introduced himself and the scrivener with whom he is obsessed, the lawyer-who-writes then makes a detour and describes his employees, his general surroundings, and his "chambers," which are orchestrated both visually and aurally. He describes himself as "an eminently *safe* man" and adds, as an aside (we are taken into his confidence, a word to be wary of in Melville), that John Jacob Astor has characterized him as "prudent" and "methodical." He "loves to repeat" the name, for it has "a rounded and orbicular sound to it, and rings like unto bullion" (4). Names, certain names with certain rich sounds, ring incessantly in the narrator's orbicular ear.

"Bartleby," in fact, will become the name the narrator most desperately needs and will therefore replace "John Jacob Astor," which itself stands in for "bullion." From this perspective, "Bartleby" is a contest between the paper of writing, legality, and accounting practices and the use of bullion as a "object" that "guarantees" a standard of value. As Ed Cutler has shown in his analysis of the nineteenth-century origins of the aesthetics of modernism,

> Bullionists blamed widening paper currency circulation as the principal cause of financial panics, warning a "shadow is not a substance." . . . legal tender advocates countered that if the bullionists were correct, then "all the paper devices of civilization, by means of which property is held or exchanged, [are] a fraud and a delusion." (2002, 31, 36)

In the social world Bartleby shadows, then, there is already a discourse of value, veracity, exchange, representation, and fraud that he will inherit, disrupt, and suspend. What if, in the world of paper on which civilization rests, he refuses to verify?

Gold and death, of course, have a long and intimate history, have always been related to a symbolic economy that has to do with incorporation and excretion, and in this narrative the orbicularity of the Astorian leads to the grim verticality of the Tombs. The fetish-thing of gold will be replaced by the system of capital exchange that depends upon a paper trail that hopes to run incessantly and without a hitch to produce profit, which will then be reincorporated into the system. Bartleby, however, is that which cannot be incorporated, even though it lies at the very center of the system of corporations, and which will therefore be excreted, in a self-defensive measure of desperation, from inside the walls of Wall Street to the apparent outside of the walls of the prison. But in the world of these premises, this chambered textuality, there is no outside/inside divide that can sustain itself. As Cutler explains: "Debates over the intrinsic or representative value of money reveal the extent to which any attempt at providing a stable representation of value within industrial modernity is itself haunted, not only by the potentially arbitrary and ephemeral nature of monetary symbolization but also by the split personality of the commodity form itself" (37). Money and the commodity are always split, and therefore spectral subjects never able to establish a foundational value within themselves.

Nonetheless, the elderly narrator of *Bartleby* continues his quest to ground his identity, for powerful factors beyond his control have long threatened his "snug retreat." "[S]ome time prior to the period at which this little history begins" (Melville 1990, 4), he had lost his lucrative position as Master in Chancery, when the new constitution suddenly and "violently" dissolved the post. Gold is good; constitutionality is bad; and the coupling of loss with violence pervades the tale.

Bartleby, the quietest of men, will provide a "sudden and violent blow, a blow that belongs not to the order of things that befall, or that can be applied, or that are real or realizable, but that has rather the inescapable persistence of a recurring dream, or of a strange whisper running around, an unattributable rumor" (Smock 1998, 96). The lawyer on Wall Street inhabits a world of intellectual premises and the trappings of wealth, but these are, in their very structures, phantasms and rumors of dreams, something that remains inscrutable to him as he attempts, under the assumption that there is an available discourse of the real, to write a "little history." This philosophy, this relationship to writing, will be shattered.

Bartleby is hired in order to keep up with the additional work entailed by the narrator's new office, and one loss accompanies the other. Our storyteller moves to his chambers, "upstairs at No. —— Wall Street" (Melville 1990, 4). He, after all, is a businessman and has things to do. Places to be. The address includes another line, a shorter one. The anonymity of a nonaddress that indicates nowhere and anywhere on that street called "Wall." *Dasein*, for this safe and prudent man, is to be in a nondeterminable place, or a place only partially determinable. Determined and nondetermined: No. —— Wall Street. Placeless. No, blank, wall, street. A chain of substitutions is constructed that moves from number to negation, and the written address displaces a certain inability on the lawyer's part to articulate a "no" to his assumptions.

The chambers, which are also denominated as the *premises* (as of a syllogism), receive a detailed description. "At one end they looked upon the white wall of the interior of a spacious skylight shaft, penetrating the building from top to bottom" (4). The shaft, which opens up to the sky high above, "penetrates" the building, and has a specific directionality. The "view"—that which is projected on the screen—is "deficient in what landscape painters call 'life'"(4). It is lifeless, already a tomb where mummies might be found walking about, doing legal work, or writing something, a little something, to make up for a loss.

Although there is no "life," there is at least a "contrast," for at "the other end of my chambers . . . my windows commanded an unobstructed view of a lofty brick wall, black by age and everlasting shade, which wall required no spyglass to bring out its lurking beauties, but, for the benefit of all nearsighted spectators, was pushed up to within ten feet of my windowpanes" (4). Everything is up close, and hallucinations—the projection of meanings—will occur where there is a topological contrast of opposites.

The interval between the walls—that which divides and connects the walls—resembles a "huge square cistern" (4). Apparently a small nothing, the interval is phenomenologically "huge," a hollowed out terrain, a depth in which to store water for daily use and in time of need. The cistern echoes with its earlier forms: the Latin *cisterna*, a reservoir, and the Greek, *kiste*, a chest or box. A cistern holds water, beer, or other liquids; is synonymous with a fountain or a lake; and, in anatomy, means a sac or cavity containing a natural fluid of the body. A "cist," in English, means: (1) a primitive tomb made of stone slabs or hollowed out of rock, and (2) In ancient Greece, a box containing the sacred tools used in the processions for the festivals celebrating the mysteries of Demeter or Dionysos. The one who carried this box, this small container of death and the gods, was called a *cistophorus*. A *cistophorus*, who carried the *cistae mysticae* in the procession, is also "a term applied to certain silver coins issued in Asia Minor, in consequence of the type with which they were impressed—a Dionysiac *cista*, out of which a serpent glides. The other side of the coin bears the name or monogram of the city of issue" (*Harper's Dictionary*).

On one side of the means of exchange, the mysteries of the god; on the other side, the city of origin. What is the between? The boundary?

Derrida tells us of a very similar structure when he is describing the archive, the inscription in the Bible, that his father passed on to Sigmund Freud as a gift:

> Arch-archive, the book was "stored" with the arch-patriarch of psychoanalysis. It was stored there in the Ark of the Covenant [Deut. 10:1–5]. *Arca*, this time in Latin, is the chest, the "ark of acacia wood," which contains the stone Tablets; but *arca* is also the cupboard, the coffin, the prison cell, or the cistern, the reservoir. (1996, 23)

Like the narrator, we are "staggered" by this "immense text" that acts as the compressed archive, the poem of the West, rippling out both backward to the primordial past as well as forward into the unforeseeable, but surely technologized, future.

The cistern, a kind of virtual mirage, is made by the lawyer's metaphorization of a space between a window and a wall and "owes" something to a structure that links the "great height" with the placement of his second-floor chambers. It preserves *both* the waters of life *and* the body of death, as well as an intermediary substance: intoxicants. Cisterns are encavements that store fluids; cist/erns are stones arranged in a rough rectangle to hold the dead. The hollows are either "natural," like lakes, or "artificial," like fountains or tombs.

This little interval between the wall of the other and "my" wall—ownership is constantly asserted—resembled the huge square cistern "not a little." The narrator "sees" the enormity and the squareness, but he does not appear to see the doubleness of his writing on Bartleby, which works to preserve life, *his own*, while arranging Bartleby in the Tomb(s) with the sacred utensils of his writing. The narrator en-graves Bartleby, who is no-body and no-thing but a virtual reality, a name generated by an arrangement of walls.

Everything in *Bartleby* happens underwater, in a tomb, in a hollow of the topography full of ghosts. There is a screen, a shaft penetrating to the sky, contrast between dark and light, an interval, and a cistern. There are fantasized "nearsighted spectators" and there is no need for a "spyglass" to enlarge the beauties "lurking" in the "everlasting shade." At each end of the chambers is a certain "loftiness"—there is the "great height" of the surrounding buildings—combined with the extremity of limits: a white wall and a black wall, between which Bartleby's story runs its short course and his refusal, which is an assertive claim of another sort, takes *place*.

The scene is open to the heights, but the *effect* of the loftiness in *these* chambers is myopia. Everything outside of the range of the "mine," here among the screens and shades of Wall Street, becomes blurred. As Bernard Terramorsi has put it: "Before it is a 'biography,' the text is a topography that defines a surface, the value and function of property: a cadastre" (1991, 90; my translation). A "cadastre" is a register of a survey of lands for the purpose of taxation, and the narrator was originally a title-hunter. The topography of the text, the way the text com- and dis-poses an architecture of the possibility of meaning and its loss, is far more essential than the "realistic" or "biographical" nature of the lawyer's recitation about Bartleby. In fact, it radically calls such categories of literature into question. Bartleby is, after all, a part of the topography—the writing of place and the writing in place—of the narrator's chambers. He has already, in name, made his textual appearance, but only now does the narrator mention the actual "advent" (Melville 1990, 4) of his coming, and, from the beginning Bartleby is walled in by the screens "astonished eyes" of the narrator, his benefactor and betrayer.

THE PHILOSOPHICAL CHAMELEON

What Bartleby *is*—to pose the classical philosophical question— has not yet been established, other than that he is an advent. We will see, however, that he is also himself a wall, to be written on by a graffitist, to be measured, to be raged against and destroyed. And, much like Nippers's table, which is treated as if it were "a perverse voluntary agent, intent on thwarting and vexing him" (Melville 1990, 7), Bartleby is also a *piece of furniture*. He is a thing that vexes and perturbs; he is a table with a will, like that of the occultists which shakes and moves, and which can only say, by pointing, yes/no.

Marx, too, speaks of a magic table. "It is as clear as noon-day"— and here, as usual, dawns the daylight of rational clarity—

> that man by his industry changes the forms of the materials furnished by nature, in such a way as to make them useful to him. The form of wood, for instance, is altered by making a table out of it. Yet, for all that, the table continues to be that common, everyday thing, wood. But, so soon as it steps forth as a commodity, it is changed into something transcendent. It not only stands with its feet on the ground, but, in relation to other commodities, it stands on its head, and

evolves out of its wooden brain grotesque ideas, far more
wonderful than if it were to start dancing of its own accord.
(Max 1978a, 320)

Nature, work, transcendence, and the grotesque are linked in the
discourse of the magical, topsy-turvy world of capitalism.[1] In this
world, the "what"—the question of essence or substance—simply
cannot be adequately answered any longer within the old table of
definitions, for its seems that "wood" can become a dancing figure,
a brain, or a person like Bartleby. Marx, like the lawyer, believes that
we can return to an authentic meaning beneath, or beyond, the illusory, through a methodology of philosophical insight that unveils
the true character of the object in question.

The narrator, still operating under the old assumptions, had a
need, and he therefore issued a call in the form of an advertisement.
He received an "answer" from a "motionless and incurably forlorn"
young man. In his turn, the narrator will be silently called by
Bartleby to write *Bartleby*, to repeat the name and its problematic,
which is profoundly linked to the history of technologics in the
form of not-literature. The lawyer who originally "draws up recon-
dite documents of all sorts" will become, passing through a crisis of
theology, a writer of a "little history."

History, or at least the desire for history as a stable narrative,
emerges as the attempt to interpret an enigma, but, try as he might,
the lawyer will never be able to understand that history, which is his
own as well as another's. The lawyer wants to domesticate, via rep-
resentational writing, the phantom that undoes all the security of
his own existence, but Bartleby is that which both calls forth writing
and refuses the possibility of an adequate writing, a writing that
might become commensurable with "life."

Upon his appearance at the door and after "touching" upon his
qualifications, the lawyer immediately "engaged" Bartleby, who was
placed among the narrator's "corps" (as employee, body, corpse, and
core) of copyists in the hope that he might act as a kind of sedative
on "fiery" Nippers and "flighty" Turkey. Bartleby the Pharmakon:
what is it that he is called upon to drug? To be the magician and
scapegoat for? What waters does he stir in the cistern?

It is from this fund that dialectics draws its philosophemes.
The *pharmakon*, without being anything in itself, always exceeds them in constituting their bottomless fund. It keeps itself forever in reserve even though it has no fundamental

profundity nor ultimate locality. We will watch it infinitely promise itself and endlessly vanish through concealed door-ways that shine like mirrors and open onto a labyrinth. It is also this store of deep background that we are calling the *pharmacy*. (Derrida 1981, 128)

Bartleby is a drug as well as the drug-store, a store that stores meaning itself, that has no "ultimate locality" and can therefore endlessly reappear and vanish.

The narrator returns to a description of his chambers, which are divided into two by "ground-glass folding doors" (Melville 1990, 9). Not opaque and not transparent, the world on Wall Street is divided into the bossman's room and the workers' room. Class always has its markers. Bartleby inhabits neither, for a new space is created for him, on the same side of the door as the lawyer, but separated from him by a "high green folding screen" (9). Folding things, which are arranged according to the narrator's "humor," regionalize the chambers, creat-ing an internal class system that divides from sight but remains voice-linked. Bartleby is a telephone circuit: "And thus, in a manner, privacy and society were conjoined" by that motionless conjunction just en-gaged by the narrator. Bartleby, who is "within easy call," is indeed projected as a "beneficial operator" in the office-economy, and he even has a window if not a room of his own.

This window, like Bartleby, is an operator that changes one thing to another, for while before it gave a view of "certain grimy back yards and bricks . . . [it] commanded at present no view at all, though it gave some light. Within three feet of the panes was a wall, and the light came down from far above, between two lofty build-ings, as from a very small opening in a dome" (9). The dirty window has been screened by "subsequent erections"—the sexual and the ar-chitectural are not unrelated—and we encounter the same language of the light "from far above" that we have seen in the previous pas-sage. The narrator wants to keep Bartleby nearby but at a certain distance; out of sight, but within calling distance.

Finally, the action picks up, and Bartleby needs food, fast. "At first, Bartleby did an extraordinary quantity of writing. As if long fam-ishing for something to copy, he seemed to gorge himself on my doc-uments. There was no pause for digestion" (9). The narrator assumes Bartleby wants copies to copy, and he feeds him his own documents. Take, eat; this is mine. Eat me. But there is no nourishment here, only intake without digestion, copying without self-reflection. Bartleby is on a mechanical binge, and the new scrivener "ran a day and night line, copying by sunlight and by candlelight. I should have been quite delighted with his application had he been cheerfully industrious. But he wrote on silently, palely, mechanically" (9).

Bartleby runs day and night; he runs a line through day and night that breaks the expected division between work and leisure, between public and private, between creation and rest. He uses the energy of the sun and of wax to light his way, but mechanically and without the good capitalist virtue of a cheerful industriousness. Capitalism will always dream of the infinite day. Already, and he's barely arrived, Bartleby has passed from being a hungry man to being a Xerox machine.

One of the tasks of the scriveners, like that of philosophers and technologists, is to "verify the accuracy of the copy" (10), and it was on the third day that Bartleby was called in from his screened retreat to do a bit of business. On this particular day, a normal day like all the other days, the supervisor was in a hurry to have his accuracy verified in the "haste of business" (9). Since time is money, business speeds up time and rushes to get things done. It is a "small," even "trivial" affair the narrator has to complete, but he is "much hurried" and in "haste." He "abruptly calls," so that Bartleby might "immediately . . . snatch" the document and "proceed to business without the least delay" (10). Employees, caught up by the *Gestell*, are always placed on call. Be available, now.

But in a gesture that threatens the entire structure of law, capital, and philosophy as *mimēsis*, Bartleby quietly refuses his place in the network: "I would prefer not to." This is not a face-to-face encounter, but only voice-to-voice, with the voice floating over or through or around the high green screen. The calm voice arrives, in the shape of a sentence that politely refuses, and "stuns" the listener. The stun gun of the language of refusal both surprises and immobilizes the narrator, and then raises his ire, which triggers the *fort* and *da* of repetition. The narrator repeats his request, and Bartleby, an echo of an echo, repeats his response.

> "'Prefer not to,'" echoed I [the narrator], rising in high excitement, and crossing the room with a stride. "What do you mean? Are you moon-struck?" (10).

Once more, Bartleby repeats his singular sentence which, eventually, sentences him to eviction and then the Tombs.

Who is copying whom now? Bartleby's refusal breaks open the patterns of expectations of business, which always responds to the needs of the now, the just-in-time. It begins a line of questioning that the narrator will not be able to let go of. The work of the negative, in all of its senses, is here fundamentally threatening all identities, and therefore the possibility of command and control. The Bartlebyian "I would prefer not to" is analogous to the Socratic "What is . . . ?"; it opens up a world. Socrates opens up the logosphere by the power of a question, while Bartleby does so by the power of negation. These are different, but not opposed, forms of language. The question negates what is assumed, the common-sensical and the apparent face value of what is before us, while the negation calls into question the same.

A rift appears, a disturbance in the chambers of the narrator's inner ear (which has been formed by a history of institutions), and he asks: "What do you mean?" What, he asks, both numbed and catalyzed by a drug, does this *sentence* of refusal mean, and what, Bartleby, do *you* mean? Are you moonstruck, a lunatic? Apparently treading the path of ethics, the narrator considers what "is best to do," and since there is nothing "ordinarily human" about his scrivener, since he is like his "pale plaster-of-Paris bust of Cicero," he returns to his desk and calls in Nippers to speedily examine the papers. He "concluded to forget the matter for the present, reserving it for [his] future leisure" (10).

The scene repeats itself with the other copyists lined up in a row, and once again, after raising the question what is wanted, Bartleby refuses to participate in the "review of accuracy" (11). This apparently trivial refusal, if passed along contagiously to others, would shatter the financial district, for "[s]ince the document and its copies supposedly embody 'truth,' any mistake or discrepancy among the copies would challenge the truthfulness underlying the whole system" (Weiner 1992, 105). Without an established system of *mimēsis* in place that more or less invisibly governs the truth, its replication, and its circulation, the economy and the law would cease to function as the well-oiled machines they are intended to be. As Jeffrey Weinstock has commented, however, there is always an

> "ineliminable residue" that remains, that exceeds the mimetic process, that accounts for the uncanniness of the double, of two things that are the same, yet are not the same. [And] in the space of the law office, this uncanniness of the double, the threat of confusion between original and copy, is contained by a strict process of *accounting*. (2004, 26)[2]

Accounting is a numerical mechanism that accounts for counting and for the veracity of the counters. It is also a synonym for the process of giving sufficient reasons for one's actions; we must, they say, "account" for ourselves.

The narrator then asks (and we knew it had to come): "*Why* do you refuse?" The *what* of philosophical description is joined by the *why* of causality. How might we understand this odd phenomenon? the narrator asks, and then proceeds to take the usual tack. But Bartleby is, so far, a speaking thing and responds to the question about cause with "I would prefer not to" (Melville 1990, 11): an answer and not an answer, for it answers why he does not want to copy—it is a question of preference—but does not give sufficient reasons for this preference. He repeats rather than explains, and the narrator, being a servant of the law, cannot afford to give up the logic of explanation. Desire, claim the lawyers and their philosophical kin, must defend itself at the bar of reason; otherwise, the sophistry of passion defeats the rationality of philosophy.

But the law itself is contaminated by an unacknowledged desire. "[S]trangely disarmed . . . but, in a wonderful manner, touched and disconcerted . . . I began to reason with him" (11). Bartleby, who is not an "ordinary human," now has a disturbing, touching, magical effect on the narrator, who responds to these currents with reasons that want to explain so as not to be touched—as a lunatic is touched, as a lover is touched—and thereby disarmed. (Soon he will be "unmanned.") Bartleby is acting as

> an "anamorphic blot" in the narrator's frame of reference. He is an inscrutable and disturbing object that should not be there and that cannot be viewed aright by the narrator except by looking awry—which would involve the distortion of the rest of the narrator's world. He is simultaneously spectral, diffuse, haunting and somehow also ultra-dense—a "black hole" of meaning. (Weinstock 2003, 32)

For now, the narrator clings desperately to reason, claiming (correctly no doubt) that it is "labor saving," that "every copyist is bound to help examine his copy," and, most importantly, it is done "according to common usage and common sense" (Melville 1990, 11). The law of the common binds Bartleby to do as he is expected to do. But Bartleby is the uncommon, that which is inscrutable to common sense and common rationality.

In a "flutelike tone"—and music, like preference, is not amenable to the entreaties of reason, as Socrates discovers as he approaches his own death—Bartleby indicates that he gave the narrator a "Yes: his decision was irreversible" (11). The Yes, as in the entire history of deconstruction from Nietzsche through Derrida, is linked with the No(t). Bartleby "never refuses to answer, he answers with infinite patience. And he always answers affirmatively. For when asked, 'Will you answer?' he does answer. Yes, his answer says. In Bartleby's mild voice *no-no answer*—says yes, yes and again yes" (Smock 1998, 75).[3]

STAGGERING

The narrator "browbeaten in some unprecedented and violently unreasonable way . . . begins to stagger in his own plainest faith" (Melville 1990, 11). Beaten down by the unprecedented—Bartleby has no original sources and refuses to rest in a tradition of copied precedents—and by the violence of the unreasonable, the old lawyer tries to balance himself by placing himself in the fantasized position of the third person and then by calling on the others to confirm his place: "Am I not right?" (12). Strange locution, dependent on the "not" and a reader's preference not to read literally: Am I—not right? Is this who I am? The not-right; the not-write? Not right in what sense? Not correct? Not virtuous? Not right in the head, a bit touched, perhaps?

Ensconced in his "hermitage"—an image that opposes the "snug retreat" of the lawyer—Bartleby continues to write without any guarantee of efficacy, as the narrator thinks more and more obsessively about the "perpetual sentry" behind the screen. He seems to live on ginger nuts, which are hot and spicy, but they have no effect on him. Bartleby neutralizes the hot: passion, anger, desire, the friction created by the hurried. And he lives by eating the metonymic equivalent of the law, for whom "the whole noble science of the law was contained in a nutshell" (8). Bartleby is devouring, bit by bit, the law of guarantees, veracity, proof, evidence, the entire system of truth. But it's a hard nut to crack, and ginger nuts, in this quantity, will act like hemlock. And where there is poison, a prison will be close at hand.

Ghosts? Keep an eye out.

The narrator takes his own measure, trying to understand the dilemma of the thing behind the screen. Passive resistance, he says to himself, aggravates the "earnest" person. "If the individual so resisted be of a not inhumane temper, and the resisting one perfectly harmless in his passivity, then, in the better moods of the former, he will endeavor charitably to construe to his imagination what proves impossible to be solved by his judgment" (13). When reason and judgment fail, he ruminates in a continuation of his Kantian mood, then one should bring in the rear guard of the imagination in order to overcome the "passive" and "perfectly harmless" resistance. Resistance, as we know, is neither perfect nor harmless.

Having done a bit of moral calculus, the narrator concludes that "Yes. Here I can cheaply purchase a delicious self-approval. To befriend Bartleby, to humor him in his strange willfulness, will cost me little or nothing, while I lay up in my soul what will eventually prove a sweet morsel for my conscience" (13). First, the narrator offered himself in the form of his documents for Bartleby to eat, and now he himself is tempted by a delicious, sweet morsel that is "cheaply purchased" and costs next to nothing. Money, virtue, and food are all entangled in the narrator's calculations. The narrator's casuistry, however, does not work, and soon he is seeking "some angry spark from him answerable to my own." (As we shall examine in greater detail when Oedipus appears on the stage, rationality is inevitably accompanied by a kind of turbocharged rage.) Things in the snug comfort of the chambers are burning up, boiling over. A conflagration is at hand, ghosts and ashes.

Having been asked to go to the post office on an errand, Bartleby says, again:

"I would prefer not to."
"You *will* not?"
"I *prefer* not."

The two worlds of discourse—let's say, Königsbergian and Viennese—strike like flints against one another and the lawyer is once again staggering about his office like a wounded man, a drunk (a cistern is used in making malt) or a lunatic.

"My blind inveteracy returned," admits the narrator. "Was there any other thing in which I could procure myself to be ignominiously repulsed by the lean, penniless wight?" (14). From a great distance, from the archaic and obsolete—by what measure?—the ghost comes at last into view as a "wight." Leading us back past Middle English, that source of common usage takes us to the Anglo-Saxon *wiht*: creature, animal, person, thing. Which? All? How are animal/person/thing related? Bartleby is being shuttled along a very strange, but predictable and traceable, chain of being. It is that hierarchical and divisive ontological grid laid out by Platonism and the Platonism of the people of which Nietzsche speaks. And, then, back up the stairs of history to wight: "an unearthly creature [Archaic] 3. a moment; an instant; a bit [Obs]." The wight un-earths. *Unheimlichkeit* and a bit of time. It's something to chew on.

What is hidden away within the wight then makes a direct appearance: "Like a very ghost, agreeably to the laws of magical invocation, at the third summons he appeared at the entrance of his hermitage" (Melville 1990, 14). The lawyer, that patron of reason and judgment, knows the laws of magic and how to summon ghosts out of the "hermitage" of his own construction. That which is hidden behind the high green screen that folds, that can be heard but not seen, comes into view, and along with it the narrator's intimation of an "unalterable purpose of some terrible retribution very close at hand" (14). Very close indeed, for the avenging angel is hovering about the "densely populated law building" (15).

The conclusion of this whole business is that Bartleby becomes a "fixed fact" and a "valuable acquisition" (15). The ghost has been petrified, at least provisionally. He continues to work, for it is not the copying itself that he refuses, but the reviewing of copy, or a kind of *proof*reading that he will not engage in. He is, for the time being, a willing participant in the disseminative procedures of *mimēsis* that the law requires, but he will not double back upon his, or another's, work to check his steps, to track down error. He will not cover his tracks. Bartleby simply does not care about error. He continues to work with "incessant industry (except when he chose to throw himself into a standing reverie behind his screen) . . . [and] *he was always there. . . ."* (15). Socrates once more floats through the air, with his standing reveries in the snow and as the one constantly there in the marketplace whom the Athenians, not able to digest, expelled in disgust.

The rhythm of the narrator is bound up with the cycles of the business day and the presence of this intruder, whom the biographer who is not a biographer—life as *bios* is not able to be written here—calls "perverse" and "unreasonable." Nonetheless, he admits that "every added repulse of this sort which I received only tended to lessen the probability of my repeating the inadvertence" (15). Bartleby, simply by refusing to play the game and sitting at *la place du mort*—the place of death and of the dummy in bridge—seems to be playing the role of analyst, helping to cure the lawyer-narrator of his anxiety attacks and his repetition compulsion. He provokes the narrator's associations and his attempt to resuture the split in his inmost chambers through the work of symbolization, but, finally, it is not a cure that sticks.

Bartleby is also the Keeper of the Fourth Key to the Chambers, where the narrator unexpectedly encounters him one fine Sunday morning while making a little detour before going to church. Placing his own key in his own lock—for don't we all presume we can unlock what is ours?—he "found it resisted by something inserted from the inside," and then the "apparition" of Bartleby appeared. The scrivener, with his "cadaverously gentlemanly *nonchalance*," comes to the door in a "strangely tattered *deshabille*," in a "dismantled condition," in a "state approaching nudity" (16). Trying to unlock his own chamber, the lawyer finds a power of resistance from that which he had installed behind the screen and which had since, unbeknownst to him, taken possession of not only a key, but of the chambers themselves during all hours but work hours, when others might come and go. The rooms are haunted, and this thing that resists, that mildly refuses to allow entrance, has the characteristics of a ghost, a cadaver, a thing (for it can be "dismantled"), and a sexual threat.

And, noted with the emphatic amazement of italics, the "utterly unsurmised appearance" (16) is strangely *nonchalant*: from the French *non* and *chaloir*, "to care for or concern oneself with." Bartleby is indifferent; he doesn't care. *Chaloir* derives from the Latin *calere*, "to be warm or ardent." "What I recognize to be living," Gaston Bachelard has mused, "living in the immediate sense—is what I recognize as being hot. Heat is the proof par excellence of substantial richness and permanence . . ." (1964, 111).[4] Bartleby, who eats spicy ginger nuts without effect and who acts as a sedative to the blazing sparks of the other gentlemen of the chambers, is neither warm nor ardent. The fires of life do not burn in him. The narrator says: thing, corps(e), but uncanny and disturbing, as if alive. Ghost: but with the question of holiness still unresolved.

Bartleby stands stone-cold where several crossroads meet, where lines are crossed and crisscrossed. Screened, he shows forth. Walled in, he breaks boundaries. Silent, he speaks. Powerless, he brings down the law. Cold, he heats things up. And the presence of the "unaccountable scrivener . . . not only disarmed [the narrator], but unmanned [him]" (Melville 1990, 16). Bartleby, whom the narrator can absolutely count on and who is absolutely accountable for his preferences, is also absolutely *un*accountable Bartleby is outside reckoning and the usual countability of *Dasein*'s existence. He is always there—that's the only thing that counts—but he is perverse, unreasonable, unaccountable, *un-*. He prefers not.

Profoundly disturbed by this not-accountable singularity, the narrator "incontinently" slinks away and is then accosted by thoughts of emptiness:

> Think of it. Of a Sunday, Wall Street is deserted as Petra, and every night of every day it is an emptiness. This building, too, which of weekdays hums with industry and life, at nightfall echoes with sheer vacancy, and all through Sunday is forlorn. And here Bartleby makes his home, sole spectator of a solitude which he has seen all populous— (17)

Wall Street is a rock upon which a church of industry and emptiness is built. Bartleby has the fourth key and brings the "sheer vacancy" of night's fall into the daylight of the law chambers, his home among the walls, screens, and furiously scribbling scriveners who verify the writing of others who have just finished copying. They are components of the machine of *mimēsis* upon which the law and the economy are based. And it is at this moment that the narrator identifies with Bartleby: "For the first time in my life a feeling of overpowering stinging melancholy seized me . . . for both Bartleby and I were sons of Adam." He admits that all these broodings are "chimeras, doubtless, of a sick and silly brain," and then says that the "scrivener's pale form appeared to me laid out, among uncaring strangers in its shivering winding sheet" (17).

Unmanned by Bartleby's near "nudity" and having made sure that nothing was "amiss"—Bartleby keeps a "bachelor's hall all by himself" (17)—the narrator falls through the melancholy of the not-there of castration and death. With "presentiments of strange discoveries hover[ing]" about him, he experiences a vision of the dead: Bartleby, with whom he has just claimed a unity-of-being. All the sons of Adam die, including lawyers who write, who would prefer to write. Death, and then: "Suddenly I was attracted to Bartleby's closed desk, the key in open sight left in the lock . . . I groped into their recesses" (17) to find Bartleby's heavy, knotted bandana that contained his savings bank. What's the secret? The key is openly in view, an invitation that offers no resistance. The locked-up is completely accessible, and it throws the narrator into a reverie about the "quiet mysteries" of Bartleby in that he "never spoke but to answer."

Bartleby is an answering machine. He offers nothing of his own; he has no original ideas and can only repeat the same sentence time and time again. He has no apparent place of origin and no relatives to speak of. But he does have a little something, and only a little something, laid by in the recesses of the desk and has an "austere reserve" about him that "positively awes" the narrator into compliance and into writing history. The reserve itself is sublime, even though the reserve protects neither Bartleby nor the narrator from anything at all, from any of the things of the world. The savings, which are necessary, do not save. In fact, they only provisionally and temporarily support a "long-continued motionlessness" and "dead-wall reveries" (18).

Frustrated, the narrator continues to brood on Bartleby and his decision to make the chambers "his constant abiding place and home." Becoming self-reflective, for Bartleby *is* that which calls for thinking, he retraces his emotional steps from "pure melancholy and sincerest pity" to the emergence of fear and repulsion. The narrator concludes, moving beyond the capacity for pity, that there is a "certain hopelessness of remedying excessive and organic ill" and that Bartleby is the "victim of innate and incurable disorder" (18) of the soul. Bartleby, the cure for the heat of Nippers's and Turkey's moods and the catalyst for the narrator's own self-examination, is himself incurable. The "excessive," which is the "innate" within the organic world, cannot be healed. If the antidote is poison to itself, there is no other that can intervene. If Socrates accepts the hemlock and the cold slowly moves from his feet toward his heart, Bartleby *is* the hemlock, cold from the beginning (in a story in which there can be no beginning), the in-between of the living and the dead.

The narrator therefore decides to ex-corporate and expel the scrivener. Bartleby prefers not to give out any data about himself. All particulars, all accidents, all contingencies are absolutely irrelevant, unimportant. It is not that Bartleby is a pure essence, a substrate, that underlies the accidents of personal history; it is just that he prefers—and this preferring cannot be measured on the grid of (un)reason—not to identify himself with, or by, those accidents. Not this, not that. No. Bartleby is the antimystic, the nonmystic par excellence, for, having (always) performed a phenomenological reduction of paring away the nonessential, it is not pure spirit or pure idea that is the remainder, but an echo against a dead-wall that can only say, as a response to having been addressed, "I prefer not to."

The origin of that sentence which condemns Bartleby cannot be located, for it does not exist as a where (or as anything else), and the lack of origin undoes the chain of causality and the possibility of the explanatory language of logic. *Bartleby* presents the wall that the program of technocapitalism, as well as the Platonic project of technologics as metaphysics, cannot penetrate, even if some light does come down from above.

When asked by the lawyer about his past, Bartleby becomes almost loquacious: "At present I prefer to give no answer" (19). It is an answer that refuses to answer; it is a giving that prefers not to give, that withholds itself in its reserve. It is not a gift taken back; it is a gift not given as an object of knowledge, but as an acknowledgment of the question. And it occurs in the "at present." It is the present tense, full of presents, becoming explicit as it emerges, as it were, from the invisibility its a priori necessity for the appearance of language and its "I." It is as if Bartleby is the "it" of the Heideggerean "there is/it gives." "The enigma is concentrated both in the 'it' or rather the *'es'* . . . which is not a thing, and in this giving that gives but without giving anything and without anyone giving anything—nothing but Being and time (which are nothing)" (Derrida 1992a, 20). Bartleby, as both text and figure, is the cipher of being and time, neither of which "is" in an ontic manner and around which writing attempts, always unsuccessfully, to organize itself in its attempt to grasp the *bios*, the *logos*, and the *technē*.

But the forlorn man's garrulousness is not just a one-shot affair either, for when the narrator "familiarly" pulls his chair behind Bartleby's screen—*he is now putting himself in Bartleby's place*—and once more, in a friendly manner pleads for him to be a "little reasonable," the scrivener's "mildly cadaverous" reply is: "At present I would prefer not to be a little reasonable" (Melville 1990, 19). He would rather not "give an answer"—answers are, after all, always calculable and, in a certain way that Socrates struggled with as well, dead ends—and he would rather not be a "little reasonable." This does not mean that he gives nothing or that he is unreasonable; on the contrary, it indicates that all the lines along which these retorts are usually organized are being rearranged.

Bartleby, as the one who provokes thought, gestures toward singularity, and as Weber has noted, "alterity intrudes in deconstructive readings not *as such* nor *in general* but rather in terms of the *singular*, the *odd*, of what does not *fit in*. And yet, what is 'singular' is not simply unique, for singularity involved here is not that of the individual but the after-effect, the *reste*, of iterability" (1996, 144). Singularity, in other words, is unthinkable without repetition, without a relationship with the common sense of the crowd, but it is not simply its opposite. It is its other, which cannot bear the pressed rush of the crowd.[5] The next day Bartleby decides upon doing no more writing. He is finished with it. Not only does he prefer not to verify the reports from others, but now he will no longer write at all. Bartleby, much to the narrator's surprise, has permanently stepped out of the drafting business of duplication and cast off writing altogether. That which provokes writing, which writes without the mimetic seduction of copying, eventually refuses, prefers not, quits—but writing, driven by the need of the narrator, continues with its own enigmatic momentum.

The narrator is at a loss at how to respond, and Bartleby becomes "still more of a fixture" (Melville 1990, 21) than before, finally turning into a "millstone" about the lawyer's neck. He is killing the poor fellow. The lawyer feels an "uneasiness," even though he also feels sorry for Bartleby, who seemed "absolutely alone in the universe. A bit of wreck in the mid-Atlantic" (21).[6] As if from nowhere, the sea swells and appears on Wall Street, and a wrecked man washes into the chambers of the law. The reservoir ripples.

That's it. The narrator has had it, and tells Bartleby, who is like the "ruined column of a temple," that he must go. Wall Street can afford neither ruins, for they speak of the future as well as the failed past, nor temples. Everything must be razed to make way for the new. The narrator leaves the nonscrivener in his office, and, upon returning down Broadway the next day mistakes a conversation about the mayoral election for a debate about Bartleby. The entirety of the narrator's experience is ordered by Bartleby's absence/presence. He is gone, nowhere, but he returns to himself with mixed emotions and tries the key in his door, "when accidentally [his] knee knocked against a panel, producing a summoning sound" (23). Once again, the spirits are summoned, and Bartleby, from inside, responds: "Not yet; I am occupied" (23).

Not yet; I am not ready. Almost, but not quite. Wait. I will come in a moment. I am not presentable. "I am occupied"; I am busy. This "I," unlike the narrator's, which has gone awandering, is occupied. Somebody is home, but this one can only be home in a space that he does not own, that belongs to a title hunter. "I was," the narrator confides in us, "thunderstruck." Bartleby, touched and luny, has already touched the narrator, who, staggering and trembling, has long ago fallen even though he seems to be strolling along in the dwindling light of a late summer afternoon. He is thunderstruck, jolted by lightning.[7]

Bartleby is the "unheard-of perplexity" that ignites the lawyer's questioning, leading him toward madness and toward the step outside of the crowd. Since he obviously, out of a sense of decency, cannot "turn the man out by actual thrusting," he considers doing some calling of his own—calling Bartleby "hard" names or calling in the police. He could walk back in and "walk straight against him as if he were air" (or a ghost) and that, certainly, would "have the appearance of a home thrust" (24). Thrust it home; fuck this guy. But, as is his wont, the narrator defers the action that would test the "doctrine of assumptions"— he is preparing to become a theologian—and decides to try once more to talk Bartleby out of it, like Freud banging away at Dora.

Bartleby prefers *not* to quit the lawyer's chambers, and that lights the fuse of his anger: "What earthly right have you to stay here? Do you pay rent? Do you pay me taxes? Or is this property yours?" (24). The theological, the legal, and the economic are conflated. How has Bartleby earned the right to remain: rent, taxes, property? "Checking" himself—he always repetitively inhibits the repetitions of his compulsiveness—the lawyer struggles to calm down. The narrator, recognizing the danger, tries to "drown" his feelings for the scrivener and is saved, at least temporarily, by the divine law to love one another. This is a "great safeguard" to its possessor, for "no man that I have ever heard of ever committed a diabolical murder for sweet charity's sake" (25). To kill for love: who, in their right mind, would do such a thing? Hoping, quite unreasonably, that the immobile would begin to move, to take on the automobility of reason, the lawyer tries "immediately to occupy [himself]" and "comfort [his] despondency" (25). As he struggles to evict the occupant of his inner chambers, he also wants his self-identity to enter and occupy the place of emptiness, for this is his only means of comfort (whether those means entail sexual thrusts in this world without women, the insertion of fantasy into the narrow office, or writing to fill the void of meaninglessness and incipient death).

When business is "driving fast" and the legal gentlemen are in dire need of help, rumors, with which the narrator cannot abide, start to swirl. Bartleby might actually live a long time, "keeping soul and body together upon his savings [as an infinite reserve which does not expend itself] . . . and in the end perhaps outlive me, and claim possession of my office by right of his perpetual occupancy" (27). As his colleagues whisper about the "apparition" in his room, the elderly writer remarks, with a kind of desperation, that "a great change was wrought in me. I resolved to gather all my faculties together and forever rid me of this intolerable incubus" (27). The idle talk of others forces a resolution.

For the moment he masters his bipolar swing, his duality that sways between hot and cold, and suggests to Bartleby the "propriety" of leaving; but the latter, after a three-day reflection, "apprised me that his original determination remained the same; in short, that he still preferred to abide with me" (27). Bartleby's original determination—not just a determined will and decisiveness, but a having been determined, particularized in his singularity, by a call from the lawyer—to abide remains undisturbed. Bartleby and the narrator are unthinkable without a repetition, but these repetitions are not the same, not identical. Bartleby remains motionless in his dead-wall reveries, while the lawyer cannot stay in place, cannot stay put.

What is to be done with this ghost? Rather than thrust the "poor, pale, passive mortal" out the door, the narrator would rather "let him live and die here, and then mason up his remains in the wall" (27). The dead-wall reveries of Bartleby have called forth an image of walled-in-death from the author.[8] He thinks of constructing a crypt in his innermost chambers where he could bury the silent dead, put his restlessness to rest. It is as if we have an example of what Abraham and Torok have described as "incorporation . . . a pathology inhibiting mourning—being responsible for the formation of the 'crypt.' The love-object (in phantasy life) is walled up or entombed and thus preserved as a bit of the outside inside the inside, kept apart from the 'normal' introjections of the Self" (Ulmer 1985, 61). The narrator—unable to mourn, haunted by the innermost figure of his own premises, and enraged by his impotence—can do nothing but construct a writing that encrypts the cryptographic enigma itself. But the code can never be broken. The advent of the Tomb(s) has appeared from what must already have been a crypt (of the law, economics, self-consciousness, etc.).

The narrator dismantles his chamber, leaving only its fixed centerpiece, and then moves into his new quarters. When the new tenant of No. — Wall Street demands from the narrator to know who Bartleby is, the latter replies: "'I am very sorry, sir . . . but, really, the man you allude to is nothing to me—he is no relation or apprentice of mine, that you should hold me responsible for him'" (Melville 1990, 28). Nothing; no relation; not responsible. The lawyer is learning to speak the "not," but only as evasion and defense, not as a phantasmic affirmation. But the "nothing" returns, for Bartleby, by doing nothing to change his situation, is now "haunting the building generally," causing an uproar. Set free from behind its private screen, the ghost who will not verify or copy drifts into public, causing panic, and panic precedes a crash. The lawyer is held to the "terrible account," which ensues when accounting and accountability fail, and returns to his "old haunt" to speak with Bartleby.

Bartleby, god of the threshold, still governs comings and goings. The narrator, flailing, now becomes a job counselor, offering careers from bartender to companion for a young man on a European voyage. He puts the whole of the American economy to Bartleby, but the latter would prefer otherwise and certainly doesn't want to set sail. He prefers "no change," to be "stationary," but he is "not particular" (Melville 1990, 30). Everybody else is particular; he is not. Bartleby wants to stay put, to be in place. But, strangely, he is not particular. Bartleby knows his place, and it is where he is. Ghosts, after all, do not wander, but stay near the place of crime and death. Bartleby is not particular, which, of course, does not mean he is a universal, either. He is not, recall, reason-able.

He is not a thing, with its dense particularity; nor is he a concept that encompasses a wealth of particularities. He is a name, but a name without a history, a name without anything but a ghost of a referent, a wisp of a trace. The name, "Bartleby," enables thing, concept, ghost, and enigma, each of which depend, first of all, on a language that names and fixes the world in place, pins it down (like, say, *points de capiton*).

Bartleby, himself, is not particular; he does not particularly care for anything and remains indifferent to the choices laid before him. He refuses to *invest* (*besetzen* is Freud's term) in his particular identity. His identity is to be without identity, *except* for his acquiescence, his yes, to placement within screens, chambers, buildings, and the urban grid of streets, prisons, and post offices. Bartleby is an occupier, a stationary place-setting around which place arises. He—although the "he" and the "it" are inseparable—is an investment who declines investment, but who, nonetheless, out of refusal and loss, enables the production of the profits of writing, even if such writing is not-literature.[9] Musing on Bartleby, Giorgio Agamben asserts that

> The perfect act of writing comes not from a power to write, but from an impotence that turns back on itself and in this way comes to itself as the pure act (which Aristotle calls agent

intellect). This is why in the Arab tradition agent intellect has the form of an angel whose name is *Qalam*, Pen, and its place is an unfathomable potentiality. Bartleby, a scribe who does not simply cease writing but 'prefers not to,' is the extreme image of this angel that writes nothing but its potentiality to not-write. (1993, 36)

The "not," finally, allows the narrator to begin to write, allows him to be summoned by things, machines, wights, and angels, even though it does not allow him to write literature.

The narrator, almost "flying into a passion," shouts at Bartleby that if he does not clear out, then "I shall feel bound—indeed, I *am* bound—to—to—to quit the premises myself!"(Melville 1990, 30). Bound in his obligation and rage, bound by the law, he begins to stammer to, *tu*, too, two, and threatens to quit the premises himself. Earlier, he had sorted through the "doctrine of assumptions" and given it up as wrongheaded, but now he threatens to quit presuming, to give up on the premise of premises. That, surely, would be the end of ratiocination.

Bartleby refuses the narrator's final offer, an invitation to go home to his "dwelling" with him, and with that the narrator "ran up Wall Street towards Broadway, and, jumping into the first omnibus, was soon removed from pursuit" (30). It was if he had seen a ghost, and for the next few days he "drove about the upper part of the town and through the suburbs in [his] rockaway; crossed over to Jersey City and Hoboken, and paid fugitive visits to Manhattanville and Astoria. In fact, [he] almost lived in [his] rockaway for a time" (30). He has become the anti-Bartleby, always on the move as a fleeing fugitive, and, just coincidentally, one of his stops is that town with a "rounded and orbicular sound" (4) that we've heard mentioned before: Astoria.

Names are once again changing places, and the Astorian, which should be the sign of the bulwark of bullion, has become only another stop on the line. The temporality of the present consists of the narrator's compulsive drive toward the past that disables him from concocting a future other than as repetition—with its disavowals and denials—and all of this will take the form of writing of not-literature and not, not quite, philosophy. He writes, it seems, a text that ghosts other genres and other traditions, or, perhaps, that others all traditions by showing the ghost at the heart of thought itself. The dead return to the present, and the present opens up upon the speaking crypt of the dead.

If his world is expanding, then Bartleby's is narrowing down to the size of his grave. He is carted away to the Tombs as, of all things, a vagrant. From the chambers of the law, he is forcibly removed to the House of Justice, a prison that is also a place of death. When the lawyer visits him, he "narrates all [he] knew" (31), already playing the role of Bartleby's historian, already practicing his lines as a writer. He finds Bartleby in a place not that different than his office cubicle, "standing all along in the quietest of the yards, his face towards a high wall, while all around, from the narrow slits of the jail windows, I thought I saw peering out upon him the eyes of murderers and thieves" (31). It is Wall Street duplicated, clarified, condensed: made into an image, but for us, not for the narrator. For, whereas he thinks he sees criminals peering out of the slits, the narrow and dark slits, he has never had this insight about the buildings that towered above his office, for those are inhabited by his colleagues.

Bartleby is not in the mood for pleasantries. "I know you . . . and I want nothing to say to you," he says. "I know where I am" (32). Bartleby, in an extremely different sense than Oedipus, is the one who knows, both where he is and where he has always been—in an inescapable prison of the law whose name is death. And he prefers, in an oddly structured syntax, "nothing to say" to him.

He has, up to now, preferred the "I would prefer not to," a refusal to act in a certain way, but not a refusal to speak. When addressed he had in the past responded, but no longer, for now the circuit of speech is broken by the inability of the narrator to stay put and to let Bartleby do the same. Language itself is falling silent, although the lawyer will cover up this silence with the chatter of his forthcoming prose. Clearly the betrayer as he passes the silver over to the grubman, the narrator will no longer be able to find the scrivener even as he goes in search of him. The office, the inner chambers, the labyrinth, the pyramid, the tomb: all are empty.

But before (and after) the death scene, there is always the question of the original, the copy, and of what will be left over, left behind. The grubman has taken Bartleby as a "gentleman forger," but the narrator assures him that he was "never socially acquainted with any forgers" (32). All his law office does is to copy originals and check for accuracy of the copy, and the narrator denies any knowledge of forgery. The narrator, after all, sees himself as a writer, and a writer, surely, is a point of origin, not a scrivener, not merely a mechanism for copying, but one in touch with the creativity of the spiritual. The narrator, deeply involved in the mechanization of human labor—and himself also part of the Xerox machine of capitalism—denies any knowledge of forgery.

A few days later the legal gentleman returns to that place "not accessible" to commoners and enters the inmost enclosure. Even more enclosed than his office, it is partitioned by folding screens, with the "surrounding walls of amazing thickness . . . [t]he heart of the eternal pyramids, it seemed, wherein, by some strange magic, through the clefts, grass seed, dropped by birds, had sprung" (33). Seed, he imagines, drifts down through the clefts, perhaps coming in spurts, and once more he is in the world of "strange magic." Grass grows, there in the pyramids, while Bartleby is dying. The ex-writer, who has refused to dine, since one doesn't live by bread alone but by ginger nuts and a place to become lost in a dead-wall reverie, still has the electric power to cause "a tingling shiver" to run through the narrator's arm and down his spine.

That is the end of the story—Bartleby has come and gone—but there is always more, an additional morsel, and the more always includes a return of that which has passed. An ending which does not end things, and a hint, but only a hint, of that which came before the beginning. It is only rumor, which may be the essential form of literature, but the lawyer is still listening for Bartleby, even after his death, trying to get the news of "who Bartleby was" in his life prior to his advent on Wall Street. Since Bartleby can no longer speak, not even to voice his preferences, the narrator attempts to speak in his stead, but he cannot occupy that space, since it must of necessity remain vacant.

The rumor—but, since it is a rumor it has already traveled widely. There is no need to repeat it here. Those dead letters are the letters of literature, of history, of philosophy. As Derrida has written, writing is "a living-dead, a reprieved corpse, a deferred life, a semblance of breath. . . . This signifier of little, this discourse that doesn't amount to much is like all ghosts, errant. . . . Wandering in the streets, he doesn't even know who he is, what his identity—if he has one—might be, what his name is . . ." (1981, 145). And, yet, mysteriously there is still water in the cistern and we still work with assiduous passion to read the smeared letters.

Bartleby is, like one reading of Being, an "utterly unsurmised appearance," an "unheard-of perplexity" that nonetheless calls for thinking, an "austere reserve" with almost nothing saved up, but which, nonetheless, "positively awes." And we find him in a hermetically sealed and publicly accessible box—with its lids, screens, keys, and secret cache of money—that is a mailbox with its own sorting mechanism: person/drug/name/thing/machine/incubus/ghost. Letters continue, as if by magic, to appear in this box and to be sent off to unpredictable locales. This is the technologics of telecommunications that both precedes and accompanies the logic of the commodity.

The box is a tomb and a sacred *kiste* in which voices may be heard, apparitions seen as they flit across the page. The page is a cistern, empty, and therefore usable as a reservoir. The finger disturbs the water and writing occurs: marks shaped by a hand; shadows duplicated, echoes resound as close as possible to the margins of philosophy without committing a copycat crime. Ghosts, dead but restless for more, for something else, for something unfinished—not yet, not quite yet. I am still occupied.

And it gives the right to reply. Or not. As we prefer.

III

The Suspension of Animation

How can ethical resistance become real—if indeed it can—before the overbearing ghostly dominion?

—Negri

7

The Drone of Technocapitalism

The marriage of reason and nightmare which has
dominated the 20th century has given birth to an ever
more ambiguous world. Across the communications
landscape move the specters of sinister technologies
and the dreams that money can buy.

—J. G. Ballard

On the occasion of his 100th birthday, Ernst Jünger briefly
commented on the century in which he has lived. Concerning
the year of his birth, 1895, Jünger recalled the Dreyfus affair
in France and Roentgen's discovery of X-rays, which "finally made
the invisible visible and made possible new measurements of the organic and the inorganic world."[1] This is where we today exist: on the
(in)calculable line between the two domains, the organic and its
other, as the ontological lines that demarcate values shift, forever
shattering. But even as the ancient grid shifts, the lines of power are
being reorganized by the flows of capital and knowledges, a social-
technical organization of the planet that "not only regulates human
interactions but also seeks directly to rule over human nature. The
object of its rule is social life in its entirety, and thus Empire presents
the paradigmatic form of biopower" (Hardt and Negri 2000, xv).

Ours is the time of the realization of cybernetics, when machines
wait on the threshold of thought and human beings are treated as

123

components of the machine world that can be cast aside, or recast into new forms, when no longer needed. In such a period, what of ethics? Can there be an ear that listens to the call of conscience if the ear is severed from a body, itself artificial, and the system of the automaton governs the possibility of the knowledge of the invisible world, the distinction between the right and the wrong, and the image of what it means to be a human being? As we are becoming posthuman, what of the *humus*, the humility of the earth? The appearance of the automaton brings with it a swarm of humming questions that concern, among other things,

* the nature of the lines between the animate and the inanimate (the living and the dead)
* how to distinguish the real from the unreal or the genuine from the artificial
* the meaning, in the Xerox age, of production, reproduction, presentation and representation, i.e. the entire range of the history of *mimēsis* (Plato is never far away)
* the question of time and the forgetfulness of time brought by the automation; its entrance en-trances both adults and children, like the hypnotic effect of staying too long in a video arcade. This is, perhaps, the most difficult aspect of the technological to think.

Jünger's 1957 novel, *The Glass Bees*, addresses all of these questions. In the novel we find ourselves listening to the threatening hum of mechanical bees and looking at dozens of severed ears that float in a pond on the estate—a walled-in "restored" Cistercian monastery within the walls of a manufacturing and design center—of Giacomo Zapparoni, the founding director of a multinational company that specializes in the cyberneticizing of film, business, and the military. As each sphere becomes another site for the colonization of informatics, they all become more and more meshed into a series of overlapping platforms. Entertainment, money-making, and the war machine become integrated into a massive cyborg in which the individual becomes enmeshed in the machinic and the nonhuman. This is not, in kind, a new development, but it is a new phase of intensification of the absorbtion of the flesh and the finite by the inorganic.

Traditional boundaries are ruptured and Captain Richard, a down-and-out ex-cavalryman, is seeking a job from Zapparoni. In the process of narrating the scene of the interview, he gives us a history of his own past and his struggle to understand the intellectual and ethical implications of his encounter with the automated bees, the cut-off ears, and Zapparoni himself, if it *is* Zapparoni himself, for there might be no distinguishable "real" entity behind the global media events that go by the name "Zapparoni"). In the "ghostly production of post-industrial capitalism," Negri writes, the "mechanisms [that produce exploitation] remain intact and become even more powerful" (1999, 10), and Zapparoni is a figure for the absolute exploitation not just of labor, but of nature and of the human itself.

If Zapparoni's ultimate goals are, as Richard muses, to "make dead matter think," to "forget time in a dreamlike trance," and to create the "philosopher's stone" by making robots that create other robots (Jünger 1960, 7–8), then he is working to realize the most primitive human fantasy of bringing the dead to life by the machinations of the "highest" of the technologies of duplication. Such an achievement would indeed be a *mysterium coniunctionis*, joining the ancient and the most new, the animate and the inanimate, the human and the mechanical, into a new order of creation that completely rearranges any previous classification of the categories of being, the most fundamental of which is that between life and death. Zapparoni is imagined, then, as a kind of technomagus, a shaman of the cybernetic world. The alchemical arts have been electrified.

At the center of this novel—for both the narrator and the reader—is *the question of the meaning of the ear*, or more literally ears, ears that have been severed from a body that itself may be human or may be artificial, if such distinctions still abide once we enter the Zapparoni Works. The ears pose one of the quintessential questions of postmodernity: what is the "genuine" or the "real" in a period dominated by simulacra? Is it still necessary, for example, to recognize the real before we can begin to discuss the artificial—and more specifically, the implications of the "prosthetic gods" that we have become or the inventive, profit-generating artificiality of all that goes under the name of Zapparoni? Does the real—which, provisionally and with recognition of the complications such terms imply, we can call the given or the natural—any longer have any ethical or ontological priority, or have we now stepped irrevocably over that enigmatic line that marked a grid of perception and value in which such terms had meaning? In which human beings, however defined, existed in a region of being distinct from, although connected with, the animal and the thing. If we have made such a move, what are the implications of that step across?

The first question concerns the line between the animate and the inanimate. In the novel the line has been already crossed over (although not yet obliterated). Zapparoni, believing nature itself to be incomplete, wants to transcend and perfect nature. This is, of course, a motif of romanticism—the desire to create a second nature—but now it is technology, and not transcendent poetics, that strives to accomplish the goal of the improvement of the natural. And the line of "improvement" will move in predictable directions, toward that which is profitable and productive. The Zapparoni Works manufactures robots that gave the impression of being "intelligent ants, distinct units working as mechanisms, that is, not at all in a purely chemical or organic fashion" (6). His first creations were tiny turtles called "selectors," which could "eliminate counterfeits" (7). One of his basic ambitions is to create self-duplicating entities, a characteristic that has traditionally been associated with "life."

A second aspect of this enigma is the link between robotics and the question of authenticity, and the "counterfeit." The issue of the counterfeit—how to distinguish the real from the unreal, the true from the apparently true—hinges on concepts such as repetition, identity, copying, difference, and self-multiplication, categories closely bound up with a Platonic interpretation of *eidos* and appearance, as well as with the functioning of the capital law on Wall Street, but it is now utterly transformed by a technological process in which, as Richard says, "matter thinks," a principle previously "operative only in dreams" (29).

Jünger, then, is exploring a dream logic that has become concretized in daily life by the conjunction of business and science, both of which claim for themselves the *ratio* of rationality. The logic of the unconscious is being externalized, as condensation and displacement take on global proportions. Automatic writing, for instance, will no longer figure, as it did for the surrealists, as the game of chance produced by the unpredictability of the unconscious; rather, it will become simply part of the program of advertising and word processing that makes the office work.

This "matter that thinks" is no longer a kind of materialist-empiricist claim for the biological basis of the human mind, but rather for the *inorganic basis of a new form of mind*: the robotic, cybernetic mind, which although originating out of human ingenuity is also simultaneously a sign of the destruction of the subject defined by any of the ontologies represented by *bios*; the *imago Dei*; the Platonic *soma* that entraps the soul of thinking; Aristotle's rational animal; Descartes's *res cogitans*; Kant's transcendental apperception without which appearances would be "merely a blind play of representations, less even than a dream" (1965, 139); and Hegel's absolute knowing of the absolute subject. And, as Heidegger has argued: "Philosophy is ending in the present age. It has found its place in the scientific attitude of socially active humanity. But the fundamental characteristic of this scientific attitude is its cybernetic, that is, technological character" (1977a, 376). Cybernetics is the science that produces and manipulates information, thus making possible the Zapparoni Works and its many contemporary analogues and rendering outmoded systems of transmission such as the eroticism of the biological.

The dream of making self-reproducing machines is, as Richard notes, the "philosopher's stone," an observation that connects the newest science with the ancient hermetic sciences, and Cornelius Agrippa is very near at hand indeed. Although Richard assumes that it is for a kind of pleasure reading along the false paths in the history of science, Zapparoni's personal library contains "early technical treatises, books on the cabala, Rosicrucianism, and alchemy" (Jünger 1960, 41). Later in the novel, Richard notes that

> The idea of plays acted by automatons was, of course, an old story; such plays had often been tried in the history of the cinema. But formerly there had never been any doubt about the automaton-character of the figures, and for that reason the experiments had been limited to the field of fairy tales and grotesqueries. Zapparoni's ambition, however, was to re-create the automaton in the old sense, the automaton of Albertus Magnus or of Regiomontanus; he wanted to create artificial people, life-sized figures which looked exactly like human beings . . . [it] was the first performance not only of a new play but of a new genre. . . . [T]hese figures did not simply imitate the human form but carried it beyond its possibilities and dimensions . . . (100)[2]

With the Zapparoni Works, the human form has been superseded, and such a move breaks open—in a violent, if quite smooth, rending—an entirely new range of possibilities, one of which is the total oblivion of humankind, whether as literal destruction or as a kind of forgetting of Being that renders us nothing but part of the machine.

There is a profound fear, in transepochal culture, of becoming incorporated into the Borg or of being attacked by the monsters spawned by technics, but, on the other hand, this is a moment of opportunity, for as Guattari argues, "A machinic assemblage, through its diverse components, extracts its consistency by crossing ontological thresholds, non-linear thresholds of irreversibility, and creative thresholds of heterogenesis and autopoeisis" (1995, 50). We *are* the aliens, we are *already* other, and the work of the *hetero-* and the *auto-* must be enacted, with as much panache as we can muster, keeping in mind that the logic of such a move must deal not with an imitation of the human form, much less an ideal Platonic form, but with a technologics of production that wills the perfection of nature along certain of its axes.

ZAPPARONI AS AUTOMATON

Shortly after meeting Zapparoni, Richard recalls the rumors suggesting that Zapparoni himself "did not exist at all but was perhaps the most cunning invention of the Zapparoni Works" (Jünger 1960, 32). The plant, the place of the production of duplicates, has produced the man rather than the other way around (much the same way that we speak of artistic works "producing" authors). "Zapparoni," in this context, would only be an effect of the means of production; "he" would be a brand name and not a personal name, a logo and not part of the *logos* that knows being, death, and the question of ethics. It is certain, however, that Zapparoni is a media event whose image is transmitted throughout the world through a "system of indirect reportage"—developed by his public relations team—that "stimulated but never quite satisfied curiosity."

The people, as fans, want to know more, always more. Such a media star, ensconced in the heavens, is "suspected of being everywhere—he seems to *multiply* himself miraculously. A person so powerful that one does not even dare speak of him becomes almost omnipresent, since he dominates our inner life. We imagine that he *overhears* our conversation and that his *eyes* rest on us in our closest and most private moments" (33; my emphasis). Through the technology of the media, Zapparoni approaches the status of a god, present everywhere, including within the inmost thoughts of his admirers. Watched with adoration, Zapparoni becomes the watcher.

God, the panopticon, and the gently tyrannical tycoon of the surveillance society become conflated. As Richard says, "All the systems which explain so precisely why the world is as it is and why it can never be otherwise, have always called forth in me the same kind of uneasiness one has when face to face with the regulations displayed under the glaring lights of a prison cell" (77). In Freud's psychic apparatus there is a censuring censor and the ego-ideal, explored most thoroughly in *Group Psychology and the Analysis of the*

Ego, that is both imitated and feared. There is always an internal police system, and disciplines of control do not respect the presumed boundary between inner and outer. All thoughts are broadcast to the public, and there remains no private interiority. This leads toward the structure of paranoia—for instance, in the case of the divine rays that speak to Senatspräsident Schreber—and I will analyze it under the name of oedipal psychotelemetry.

Impressed by the charismatic bearing of Zapparoni, Richard concludes that he was not an "impersonator" (61), but soon thereafter comments that his eyes

> were extremely powerful. . . . The impression was slightly artificial, as if it resulted from some delicate operation. . . . This was not the blue of the sky, not the blue of the sea, nor the blue of precious stones—it was a synthetic blue, fabricated in remote places by a master artist who wished to excel nature . . . . His look cut like a blade of flexible steel. (63)

Artificial, operation, synthetic, excelling nature, and cutting like steel: such is the language, perhaps only indicative of Richard's hyperbolic imagination, used of his first encounter with Zapparoni's gaze, and it is a language in which the eyes become the synecdoche for the entirety of cyborg culture.

The man himself, however, is nothing like the grandfatherly image projected around the world, so Richard concludes that

> Zapparoni must certainly have had a deputy to play this role, perhaps an actor, perhaps a robot. [What's the difference?] It was even possible that he employed several such shadows or projections. This is one of mankind's ancient dreams, and has given rise to special turns of phase: "I cannot be in four places at once," for instance. Evidently Zapparoni not only believed it to be possible, but considered the divisions a profitable extension and intensification of his personality. Now that we are able to enter apparatuses and leave part of ourself within them—for instance, our voice and our image—we enjoy certain advantages of the antique slave system without its drawbacks. (64)

Richard, a technical-military man with a respect for tradition, here articulates several important aspects of the spectral logic of technology.

It speaks to the "ancient dream" of a self-division that is simultaneously a self-multiplication. Division, in this register of the imaginary, is not a lessening of the whole—a formulaic proposition restated when we find the ears floating in the pond and restated who with the question of amputation/prosthesis arises—but an increase, an "extension." This *holographic imagination*, in which the whole is projected into innumerable parts without losing itself, is a very old form of imagination that can be traced back through the Renaissance micro/macrocosmos (including its alchemical variants) to the Stoics' *logos spermatikos*, and all the other perspectives based in the idea of the continuities of natural law (whether of ethics or epistemology) in which the whole is reflected in every part. Holographicity in its traditional sense is a form of animism, enlivened by a divine spirit that binds the manifold into a cosmos intelligible to *Dasein* with its body, moods, and understanding. *Holon* is Greek for "whole," a term with a richly permutated history. For Aristotle, as F. E. Peters explains,

> *eidos* of living beings and the unitive cause of all their functions is the *psyche*. In this fashion parts (*mere*) are transformed, by the notion of function, into organs (*organa*). An organ is the part of a living creature that is directed toward an end or purpose that is an activity; nature (*physis*), the internal principle of growth in these beings, has made the organs to perform certain functions, and a body so constituted is an organism. The *organon*, then, is the physical part of a living being matched to each of the latter's potencies to enable them to function. (1967, 86)

The Zapparonian form of holography, governed by a different technical-economic matrix than existed in classical Greece, is radically different than the traditionalist writing of the whole. In Aristotle "parts" are transformed within living beings into "organs," which have their own separate teleological functions and work together to make up for an "organism." The Zapparoni Works dismantles that constellation of understanding, and what were once organs are fabricated into parts: uniform, mass-produced, replaceable. The body without organs becomes literalized and commodified.[3] Function alone remains, and comes to dominate all ideas of *eidos*, *psyche*, or *organon*, replacing these with its own *telos* of profit and power. *Physis*, nature's self-blossoming, is replaced by *technē*, the production processes of cyberneticized systems that will correct nature's flaws and produce, in the end, a second nature.

Not only can parts be manufactured to replace the fallible and finite body produced by nature, but the "natural" organs—voice and image, for example, in the telecommunications industry—can be detached from the holistic organism and be stored, reproduced, sold, or used for leverage, all apart from the will of the presumed original. "I"—as part-object that contains, if not the whole, then at least the authenticating marks of the original—can be multiplied as often as desired, but this is a desire that can be located at any number of points along the production system, not just within the traditional subject to whom the voice and image once belonged. "Everyone can dream," writes Baudrillard,

> and must have dreamed his whole life, of a perfect duplication or multiplication of his being, but such copies only have the power of dreams, and are destroyed when one attempts to force the dream into the real. The same is true of the (primal) scene of seduction: it only functions when it is phantasmed, reremembered, never real. It belonged to our era to wish to exorcise this phantasm like the others, that is to say to want to realize, materialize it in flesh and bone, and, in a completely contrary way, to change the game of the double from a subtle exchange of death with the Other into the eternity of the Same. (1994, 95)

The dream-cast is now. Flesh and bone, mediated by the precision of instrumentation, method, and number. Immaculate conceptions devoid of sweat, anxiety, the spread of fluids, hair, the mouth, and the closeness of the foreignness of the other.

The primal scene is the scene of cloning, displaced as if in a dream that moves among us, and the wish is the temporal and spatial multiplication of the self beyond the painful limits of finitude. To what extent the artificial has replaced the natural in the figure of Zapparoni is not precisely knowable, but the principles are being established that will allow for a completely artificial intelligence to stand in for what nature had taken millennia to produce. Things are accelerating.

THE HUMMING OF THE BEES AND THE DISCONNECTED EAR

Zapparoni asks Richard to go and wait for him in the garden, and warns him "Beware of the bees!" (Jünger 1960, 84). As soon as Richard enters the domain, which he likens to a danger "zone" rather than to a traditional garden, time begins to "run faster," and he feels it was "necessary to be more on guard" (85). The presence of the technological increases the tempo of time; it forces the biological, slow and old-fashioned as it is, to try to keep up with the pace it sets. Richard watches artificial bees of various sorts working the flowers and remarks that "Zapparoni, that devilish fellow, had once gain trespassed on nature, or rather, had contrived to improve nature's imperfections by shortening and accelerating its working methods" (93).

The fabricated bees are also more "economical" than nature's variety, and are able to drain the flower more completely. The new hives are like "automatic telephone exchanges" and the "entrances functioned rather like the apertures in a slot machine or the holes in a switchboard" (94). The natural procedure had been "simplified, cut short, and standardized. . . . From the very beginning he had included in his plan neither males nor females, neither mothers nor nurses" (95). Zapparoni's Bees, Incorporated, which "radiated a flawless but entirely unerotic perfection," is a completely celibate, although highly productive, machine. Any sexual arrangement, involving as it does wasteful effort, has been eliminated for the sake of higher productivity and greater functional efficiency. The timed utilitarianism of Taylorism governs the production schedules of the newly fabricated natural.

Glancing around after an hour's observation of the system at work, Richard notices the Smoky-Grays, automatic bees who are acting as foremen or observers of the others; and then he remembers Zapparoni as the "invisible master" who controls this "dance of the spirit, which cannot be grasped by calculation" (96), but only guided by science as a "call of destiny" (96). This, however, can only be such a call if the call comes in the form of a demand for efficiency, speed, functionalism, and the perfection of nature along the axes of these technical aspects. And it is only when the imaginative response to the call comes in the form of profitable calculation that the automatons will be produced as a self-regulating system in which there are, as Marx saw, "masters of surveillance" at work in the feedback loop. Even the "invisible master" will, sooner or later, be swept along, unable to exist without his creations, in the technical version of the master-slave dialectic.

The erotic has been rendered superfluous. After all, Richard thinks, "Bees are not just workers in a honey factory. Ignoring their self-sufficiency for a moment, their work—far beyond its tangible utility—plays an important part in the cosmic plan. As messengers of love, their duty is to pollinate, to fertilize the flowers" (98).[4] Eventually, the "true bees" will become extinct and give way to their mechanical descendants, which, while they are the victors in the contest of honey production, are no longer messengers of love. And if the flowers are not pollinated, industry will have to create better simulacra of flowers as well. Both perfumes and tastes will be synthesized; the smell of hazelnut or lavender will waft through the underground halls of the metro.

Richard considers the likelihood of the use of such microrobots as weaponry—and this is certainly no longer a fantasy—then realizes that the bees, and their "progeny," were not being designed to be an individual "commodity," an individual thing for exchange, but a prototype for a new system of energy conversion and storage that aimed at delivering "power." The bees, at this point, represent the *Gestell* itself, which governs the production of all commodities, including the human resources at its beck and call, but is not a commodity itself. The essence of technology is not itself technological, but rather "starts man upon the way of that revealing through which the real everywhere, more or less distinctly, becomes standing-reserve" (Heidegger 1997b, 24). What was already present in Bartleby and Marx, as indeed with Plato, comes to a kind of robotic fruition in the transepochal, the period of the irreversible crossing.

"The air," Richard says, "was now filled with a high-pitched, uniform whistling sound, which, if not exactly soporific, at least blurred my perception; it was not unlike the effect produced by hypnosis. I had to make an effort to distinguish between dream and reality in order not to succumb to visions which spun out Zapparoni's theme on their own" (Jünger 1960, 105). The phantasmagoria begins to whirl at faster and faster speeds, until Richard can no longer "keep up with the task of interpretation" (105) and he "starts dreaming; images get hold [of him]" (108). Before he can even identify—before the senses can correlate through some form of schema with the *logos* of meaning—Richard suddenly rubs his eyes and decides that he has been deceived in the "garden where the diminutive became large," in this place that distorts perception because technology is at work and exaggeratedly augments human sensation from the micro to the macro and vice versa. Everything is magnified. At the same moment, Richard observes, "I heard inside me a shrill signal like that of an alarm clock, like the warning signal of a car approaching with brutal speed. I must have seen something prohibited, something vile" (109). Here in the Garden of Good and Evil, he has come upon the obscene, and there is the experience of an internalized voice of warning—something like and unlike the Socratic *daimon* or the Freudian superego—but it is a mechanical voice in a sense that neither of the other examples could be, even if Freud is heading in this direction when he names the psyche an "apparatus." The warning is "shrill," like an "alarm clock," and sounds as if it comes from a car bearing down with "brutal speed."[5]

The machine, which has been internalized and transformed into the metaphorical basis for an early warning system, sounds the alarm and announces that Richard *had already seen* something prohibited and obscene. He had just said to himself that "the sundew *is*, after all, a carnivorous plant, a cannibal plant" (109), but he had not recognized, he had not interpreted correctly, what he had perceived. Perception had outraced judgment, even though a part of his perceptual apparatus—the unconscious is as good as name as any—*had* seen what was there: a human ear cut from the absent body with "neat precision" (110).

Richard then realizes that this is the moment when he "ought to speak of morality" (111). He reflects on the "mutilated body" as a specifically modern phenomenon, arguing that "the increase in amputations is one of the indications of the triumph of a dissecting mentality. The loss occurred before it was visibly taken into account." Richard concludes with the observation that

> Human perfection and technical perfection are incompatible. Technological perfection strives towards the calculable, human perfection towards the incalculable. Perfect mechanisms, around which stands an uncanny but fascinating halo of brilliance, evoke both fear and a Titanic pride, which will be humbled not by insight but only by catastrophe. The fear and enthusiasm we experience at the sight of perfect mechanisms are in exact contrast to the happiness we feel at the sight of a perfect work of art. We sense an attack on our integrity, on our wholeness. That arms and legs are lost or harmed is not yet the greatest danger. (113)

The perfect mechanism inspires fear and "Titanic pride," while the perfect work of art inspires "happiness." This opposition between the machine and art needs interrogating, but the most immediate conclusion to be drawn is that the cultural work of technology is analogous to the way in which Richard perceived something unconsciously *before* he was able to correctly identify the cause of his reaction and then begin to rationally explicate the obscene, initially prohibited, object.

Paul Virilio, speaking of science fiction, a genre in which *The Glass Bees* could easily be included, notes

the various levels of a certain anesthesia in our consciousness that, at every moment, inclines us to see-saw into more
or less extensive absences, more or less serious, even to pro-
voke by various means instantaneous immersion in other
worlds, parallel worlds, interstitial, bifurcating, right up to
that black hole, which would be only an excess of speed in
these kinds of crossing, a pure phenomenon of speed, abro-
gating the initial separation between day and night . . .
(1991, 77)

Speed is a drug and it is always too late. The day and night line of
Bartleby merges the "initial separation," allowing work to hasten
on its way, regardless of the old rhythms of the celestial or the
circulatory system.

Richard realizes that the scene, which is a kind of primal scene
of the technological age, "led to a lower level of reality . . . [and that]
everything might have been a mirage" (Jünger 1960, 110), remarks
that reveal that even as he is attempting to decipher the meaning of
Zapparoni's works, he does so in the category of traditional metaphysics, which marks off the lower from the higher as the negative
from the positive. Even if the value of the good within a Platonist-
humanist tradition is in the process of being destroyed, it nonetheless continues to act, as it limps along, as a standard-bearer, a
measurement, an instrumental grid that gauges the direction and
rate of descent into the netherworld. And, with Richard's comment
that the scene "might have been a mirage" we are within the world
of appearances in which the appearance might still be distinguished
from the reality: where, that is, the force of Platonic rationality still
operates as a means of making distinctions between the true and the
false, the real and the illusory.

Confronted with something new that springs from a very ancient dream, Richard can initially only respond to the uncanny appearances with traditional means of interpretation, even if in
some region of his being he recognizes that his understanding,
using as it does a recognizable system of evaluation, cannot keep
up with a reality which outpaces all thought. As he has already
said about the swarm of mechanical bees, he "couldn't keep up
with the task of interpretation" (105). If this is always the case,
and even more so now that technology with its "brutal speed" drives the production of meaning, there seems to be nothing to stand
in the way of the self-realization of the technical subject, which
has become an object.

Richard counts two to three dozen ears, and, trying to calm himself, declares that these ears "must have been a *delusion*; I must have been the victim of a *vision*. The air was sultry; the garden seemed to be *bewitched* and the swirl of automatons had *intoxicated* me" (136; my emphases). Richard does not trust his own perceptions and reverts again to a list of predictable moves when faced with the unknown, or only partially known: the ambiguous. His sight of the ears must have been "delusion," "vision," "bewitched," or "intoxicated." His perception of several dozen mutilated ears could only be understood under the categories of mental aberration, hallucination or trance, an evil magic, or intoxicants that impaired his rationality.

It could not be possible that there are in fact human ears floating in this pond. Therefore the explanation must lie elsewhere. The inexplicable is, from some vantage point, explicable. Rationality is still functioning as Richard seeks an explanation for a bizarre phenomenon that does not fit his experience, but it is a rationality of habit running on autocontrol. Simply shuttling along the traditional axis of interpretation, the first place Richard seeks an explanation is in the opposite of individual rationality: in the subject's own irrationality.

He next asks whether, in fact, the ears are genuine. After all, the garden belongs to a kind of counterfeiter, one who produces reproductions that are as if real, all for the sake of knowledge, profit, and the transition to the order of the new Titans. Richard examines them more closely with his binoculars, a set of surrogate eyes that brings the distant close:

> [T]he objects were infernally well done—I might almost say that they surpassed reality. . . . A big blue fly descended on one of these shapes, a fly like those one used to see around butcher shops. . . . The fly? The work of art was to all appearance so perfect that not only my eyes but the insect itself was deceived. It is generally known that the birds pecked at the grapes painted by Apelles. . . . Moreover, who in the garden would swear an oath that this was natural, that artificial? (137)[6]

It is *at this moment* that Richard enters into the transepochal labyrinth of perception and judgment, and he realizes that an artificial bee implies the possibility of an artificial flower, ear, fly, and whatever else might come his way. The artifactuality of the artifact might be found anywhere in what used to be nature, and in either the part or, at least eventually, the whole. Turtles, ants, bees, puppets, ears, limbs . . . what's next? Richard remarks

> [D]uring this strenuous testing and watching, I had lost the capacity of distinguishing between the natural and the artificial. I became skeptical of individual objects, and, in general, I separated imperfectly what was within and what without, what landscape and what imagination. The layers, close one upon the other, shifted their colors, merged their content, their meaning. (138)

Everything is merged: con-fused. The normal functioning of the "senses," as well as the transcendental "sense" that joins the manifold of the senses to make sense of the world and allows us to distinguish the inside from the inside, the imaginative from the empirical, has shattered in a kind of synesthesia (or syn-anesthesia). How else to respond to trauma?

The fundamental differences of the world, whereby we make distinctions between "domains of being," have exploded, leaving Richard in a hall of fragmented mirrors, one of which reflects his own stunned face that is. He felt perhaps, he is beginning to look like a puppet pulled by invisible strings. But, as the line between the organic and the inorganic shifts, when humans begin to become puppets, so, too, "The marionettes became human and stepped into life . . . I saw the entrance to a painless world. Whoever passed into it was protected against the ravages of time" (138). Here in the new garden of artificial marionettes, puppets with human capabilities, and ears that littered a pond, Richard concludes that "[o]f course they were artificial—or artificially natural [*natürlich künstlich oder künstlich natürlich*]—and, as with marionettes, pain becomes meaningless" (138). The reversibility of noun and adjective indicates a deep ambiguity about the substance of the thing in question: which is primary, art or nature, the made or the given? Both, in the cybernetic garden of danger, are enlivened, animated, and capricious. But only one, at least for the continuing moment, feels awe, pain, and the "ravages of time."

It is *pain*, then—a product of the organic and temporality—that becomes the possible measurement, perhaps the last indicator, of a specifically human reality, of genuine existence, and pain is fundamentally connected to the question of time that, until now at least, only *Dasein* has known. When the descendants of either Pinocchio or of Zapparoni's metallic bees come "alive," they will not enter into the pain of mortality. Destruction, assuredly, will continue to occur, but not death. When *Dasein* steps the other way, into the safety of the automaton, pain will begin to recede, and when death's sting, at long last, vanishes, *Dasein* will have ceased to exist. There will no longer be Being, there. "The threat to man," Heidegger argues, "does not come in the first instance from the potentially lethal machines and apparatus of technology. The actual threat has already affected man in his essence. The rule of enframing threatens man with the possibility that it could be denied to him to enter into a more original revealing and hence to experience the call of a more primal truth" (Heidegger 1977b, 28). The rule of technologics as metaphysics *shuts down* the emergence of a more primordial relationship to what-is and what brings beings to pass.

Still conscious that he is on a job interview and wanting to make a good impression, Richard pulls an ear from the pond, admitting to himself that the "replica was excellent. The artist had gone so far in his naturalism that he had not even forgotten the tiny tuft of hair, characteristic of the ear of a mature man, which is generally trimmed with a razor blade. He had, moreover, indicated a small scar—a romantic touch" (140). He realizes that his perceptions had all organized around the word "hear" and then thinks that in Zapparoni's Garden of Delights,

> a mind was at work to negate the image of a free and intact man. The same mind had devised this insult: it intended to rely on manpower in the same way that it had relied on horsepower. It wanted units to be equal and divisible, and for that purpose man had to be destroyed as the horse had already been destroyed. (Jünger 1960, 141)

The horse, as representative of a long history of the coexistence of the human and the animal worlds, has been replaced by the tank—Richard has been both a cavalryman and a tank inspector—and now the tank is on the verge of vanishing, to be replaced by the deadly hum of predatory drones. When such lines are crossed, humanity, too, becomes a standing-reserve to be divided, remanufactured, and cast off as soon as it becomes obsolete.

The presence of the severed ears raises all of these questions, and Zapparoni, of course, has a story to tell, an explanation to proffer. One Signor Damico, a Neapolitan, had been in charge of creating the ears for all of Zapparoni's life-sized marionettes, such as

Romeo and Juliet; it was a process that "rested less on the faithful reproductions of their bodies than on deliberate deviations . . . since one ear is unlike another." Along these lines, the public had to be taught a "higher anatomy" (145). With this pedagogy another turn in the history of mimeotechnics has been taken, an *Aufhebung* of technical artistry: Zapparoni and his minions have stepped beyond the standardization of machine culture and into the principle of difference as a copyable phenomenon. (Plato's dilemma in the *Cratylus* has at last been answered.)

Zapparoni has understood that difference, and not sameness, is the secret to the higher anatomy—not only of nature's creation of human beings but of our own creation of the next type of being, the animated marionette; but, in addition, he has realized that difference can be manufactured and drawn along in the wake of the project of technical perfection. Difference can be, as it were, tamed and subjugated, and employed in the service of the same. This is the point of greatest danger. Out of petty jealousy Damico, the creative master of the differentiated earlobe, had angrily sliced off all the ears from the marionettes he had been working on and had left Zapparoni's employ for other work. The ears, Zapparoni explains, had not been "simply stitched on or manufactured by the piece, as a wood carver, a sculptor, or a wax molder would do. On the contrary, they must be organically joined to the body by a method that belongs among the secrets of the new-style marionettes. . . . And to marionettes of this type you could not simply attach a severed ear, any more than you could to real human beings" (145–46).

Once again, as with the identity of Zapparoni himself, we find ourselves straddling the line between the organic and the inorganic, the animate and the inanimate. There is an "organic" connection between ear and body, as in the case with "real human beings," but, nonetheless, the designed and fabricated nature of the ears is emphasized as well. The Aristotelian metaphysics of noncontradiction—either the organic or the inorganic—has given way to the uncanny logic of both/and.

Listening for Nuances through the Technologized Ear

As Richard reflects on the honey-gathering activity of a glass bee, which is purely mechanical and doesn't pollinate the flower in return for its gift, he realizes that "On this level economic considerations were entirely unimportant; here one had to enter into another sphere

of economy—the titanic. One had to make a different accounting"
(101). It is this *different accounting*, in every sense, that we will have
to begin to account for as the old lines of value continue to shift at a
dizzying rate of speed. Marcus Bullock articulates the dilemma in
which we find ourselves when both explanation and inquiry—as either traditional philosophy or literature—are outpaced by the pace of
the technological. Speaking of how Jünger's understanding, by the
period of the *Glass Bees*, had surpassed that of his earlier essay, *Der
Arbeiter* (1932), Bullock notes that

> To mobilize the power of machines, the *Arbeiter* had to imitate
> them: now Jünger sees the consequences of that imitation. By
> coming to resemble what we are not, we cease to be what we
> are. People make themselves resemble machines, machines
> become indistinct from people, and people are valued and uti-
> lized in the same way as any other mechanism. There is no
> longer anything sovereign or sacred about the image of a man.
> It may be dismembered without qualms, and the isolated
> parts, or the mimicry of such parts, may be contemplated with
> equanimity. (1992, 168)

Which leaves us in something of a quandary, for we now find
ourselves in a state of suspended animation. As Richard recalls a
friend who leapt to his death, he remarks, "I had a feeling that like
Lorenz we were all jumping out of windows, and sooner or later we
were bound to crash. At the moment we were, so to speak, *suspended*
in mid-air" (Jünger 1960, 57).

The suspense can be killing, but it is also our contemporary
form of life. The old grid is suspended—we have to wait to find
out what will happen at the end of the story; a rule has been broken and play is temporarily suspended—but the suspension also,
like the cables on a bridge, keeps us moving along a road (although this road is not a *Holzweg*, but an express-way or an In-
fobahn that cuts through cities and countries alike). And, as
Guattari has rightly insisted, "Just as scientific machines constantly modify our cosmic frontiers, so do the machines of desire and aesthetic creation. As such, they hold an eminent place
within assemblages of subjectivation, themselves called to re-
lieve our old social machines which are incapable of keeping up
with the efflorescence of machinic revolutions that shatter our
epoch" (1995, 54). Shattered, but retaining the possibility of
desire and aesthetics, at least for a while.

We must, at least for the time being, continue to ask the questions of ethics in this period of the digitization of the world and the growing power of the cyberneticizing sciences. These are old questions: what shall we do? What is the good decision and the good life? What is the nature of human being? But we must also recognize that such questions are radically threatened, if not already superseded, in an epoch when the *Gestell* sets us up ever more rigidly in a technological webwork that is not only beyond the control of any one person or collectivity, but that, in some ways, makes all individualities and collectivities *effects* of the web. By the end of the novel Richard has accepted a position at Zapparoni's Works as part of the "internal arbitration system," and has become part of the industry of making automatons, puppets, and virtual realities. While Bartleby prefers the not, Captain Richard says yes, I'll accept. As have we all.

Which is to ask once more: what do we hear with our ears, both inner and outer, and what does it mean that severed ears are floating on a pond in the midst of Zapparoni's, and Jünger's, *Werke*?[7] Thinking, as the interpretive quest for understanding or as the ground for ethics, will never be able to "catch up" with the brutal speed of technology. We cannot "solve" the suspension of the animate, the destruction of the old ontology, for it is not a problem to be solved—although it produces innumerable problems that we must attempt at all costs to solve—but, rather, a phenomenon that is not amenable to instrumental reason. The essence of technology is only open to a different thinking, a different accounting-for.

In response to the catastrophe—which has always been with us, but now appears in a cyberneticized form—we can keep the question, and questioning, alive. We can pass the question on to one another; we can establish sites between us where the question can address us, where we can learn, bit by bit, to hear the question more deeply. Heidegger gestures toward this possibility: "Always unconcealment of that which is goes upon a way of revealing. Always the destining of revealing holds complete sway over man. But that destining is never a fate that compels. For man becomes truly free only insofar as he belongs to the realm of destining and so becomes one who *listens and hears* . . ." (Heidegger 1977b, 25; my emphasis). We will return to this hearing, and the timing of hearing, but whether it will turn toward us is not predictable.

Perhaps we might learn something about hearing aids and learn to hear *through* the technological—with the help of the technological, as it were—all those nuances of human existence that are threatened by the almost silent noise of the flow of the incessant computerized enframing in which we find ourselves here at the turn of the millenium. For as Jünger says, speaking through a fictional mouthpiece set firmly in the circuitry of the mediatized system of telecommunications: "There are nuances, now hardly distinguishable, but nuances frequently make all the difference" (1960, 118). We all must closely attend to such nuances, with an ear bent close to the earth and an eye on the satellites circling like tiny silver points in the night sky above.

8

The Psychotelemetry of Surveillance

> The madness of Oedipus has become Western reason.
>
> —J.-J. Goux

(Post)Modernity and Freud's Construction of the Oedipus

If Zapparoni represents the possibility of absolute surveillance in a world dominated by the electronic surveillance produced by the twinning of science and capitalism, his domain has been under preparation, as we have seen, for a very long time, passing along a network articulated first by Plato but carried along by all the transformers in the West. That net included Oedipus, which also creates its own relays, the best known of whom is Sigmund Freud, who takes it on as an identity-forming complex. Everyone, without thinking, knows what "Oedipus" represents in the work of Freud. That old story about the violence of misrecognitions is the kernel of psychoanalysis and therefore that without which psychoanalysis is unthinkable. But everyone is not every-*one*, and a singularizing process produced by

143

(re)reading must occur if we are not to be simply carried along by the ready-made-ness of what we thoughtlessly call "Oedipus."

Reading is that form of material resistance that slows us down, and, through very hesitant steps, attempts to retrace a short passage of the network activated by the multiple transfers Freud accomplishes. As Samuel Weber has reminded us, "the 'net' at work in Freud's text is not a stable or distinct *object* but a *movement* in which one becomes entangled, *verstrickt*" (1982, 76). We are already netted, caught in an oedipal circuitry that insists we seek to understand what has already happened and will happen again tomorrow,

The Oedipus—as title, name, example, complex—is a riddle. The orality of legend, the writing and performance of Athenian tragedy, translations across languages and cultures, and the technical usage of the term within psychoanalysis: all of these transfers, and uncountable others, have already, and with a hypnotic speed we cannot have followed, set the stage for our understanding of Freud's legacy. Oedipus, I suggest, is Freud's most resonant name for a form of *caulking*, that which fills the seams and gaps so that a vessel—a book, a subject, psychoanalysis—can have a stable boundary and thereby stay afloat. Etymologically, along one of its branches it goes back to L. *calx*, "heel," and through another to the OHG "squinty, crooked." Then off (simply following directions) to "cylinder" and "crime, wickedness" and then to Alb. *tschalë*: "lame; basic meaning: turning, bending." Caulking, in all of its grammatical usages, is the common noun that is Oedipus's less well known name.

The story, as everyone knows, opens outside the pages of the Dream Book. Freud's first mention of Oedipus, the *tyrannos* (also his name for psychology), emerges in the midst of his own self-analysis, the results of which he shares in the middle of October 1897 with his nose-and-sex man, Wilhelm Fliess:

> If that is the case [love of the mother and jealousy of the father], the gripping power of the *Oedipus Rex*, in spite of all the rational objections to the inexorable fate that the story presupposes, becomes intelligible, and one can understand why later fate dramas were such failures. Our feelings rise against any arbitrary, individual fate such as shown in the *Ahnfrau*, etc., but the Greek myth seizes on a compulsion which everyone recognizes because he has felt traces of it in himself. Every member of the audience was once a budding Oedipus in phantasy, and this dream-fulfillment played out in reality causes everyone to recoil in horror, with the full

measure of repression which separates his infantile from his present state. (1954, 224)

This is the first, compressed version of material that will, almost like the analysis of a dream or a symptom, be considerably expanded in *The Interpretation of Dreams*. Freud's presentation already exhibits many of the characteristics of the apparatus that would play such a nuclear role (as constitution, diagnosis, and dissolution of a historically overdetermined form of subjectivity) in psychoanalysis.

What Freud does with the oedipal material, that material which is explicitly named after Oedipus (and not after Hamlet or Cain) seems quite simple—it is all spelled out in his correspondence with Fliess—but the "simple" is always an idealized object that entails a having-been smoothed out, a sanding of the jagged shards of history. (What will the language become as archaeology is replaced by the black matte finish of the electronic archive?) Immediately before introducing the oedipal material in *The Interpretation of Dreams*, Freud analyzes the situation of an obsessive man whose father had recently died. "A person, [the patient] felt, who was capable of wanting to push his own father over a precipice from the top of a mountain was not to be trusted to respect the lives of those less closely related to him; he was quite right to shut himself up in his room" (1969, 294). When Freud again mentions this patient, it is in the context of his "Cain-phantasy." Hellenism and Hebraism meet in Freud, who moves from Cain to Oedipus as he moves from Fliess to the *Interpretation*, with Moses waiting in the wings. The patient was thirty-one years old, the same age as Freud (also a mountain climber and a patricide of the old order) when he met Fliess. After Oedipus, in both the letter and the *Interpretation*, he moves immediately to Hamlet. I cannot analyze this material here, but do want to note Freud's comment about the "two widely separated epochs of civilization: the secular advance of repression in the emotional life of mankind" (298). He recognizes the differences entailed by history, but, nonetheless, Oedipus spans historical time. This raises important questions about Freud's notion of history, the timelessness of the unconscious, the universality of Oedipus, etc.

In the short, interminable passage in *The Interpretation of Dreams* where Freud first publicly analyzes what will come to be called the Oedipus complex, the kernel of psychoanalysis, how does he handle the material? What, indeed, *is* the material that Freud is handling? *What*, in other words, is the Oedipus? It is a many-faceted title and proper name (of what? given by whom?) whose source is not clear. As we once again begin to (re)read the Dream Book, rifts, proliferations, and ambiguities appear. This material is undoubtedly Greek—Sophocles is the prototypical Athenian—but it is also universal (*allgemeingültig*). It is, simultaneously, non-Greek. It is both particular and generalizable, although the generalizable will carry the particular along to the far corners of the world—in, for instance, the attempt to "prove" the universal nature of Oedipus—and, in another sense, will bring the world "home" to the technical legacy of the rationality of the West.

Goux, for example, has argued that Oedipus provides the fundamental opening of western rationality and that "avoidance of initiation is also a liberation. It opens up a new horizon. It defines a subject who, in his disruption and his disproportionality, can live out other possibilities. The adventure of Oedipus inaugurates the era of the hero whose identity is not defined by a tradition and a transmission, a hero with whom a new mode of subjectivity emerges" (1993, 82). And this "new mode of subjectivity" of the one who would know will bring a certain sense of time, method, tone, and epistemology along with it.

"Oedipus," as a captivating structure and method, will become one name, the primary psychoanalytic name, for the shuttle of cultures that Freud, ignoring for a moment the difficulties of translation between localities, weaves between a mythologized Thebes, fifth-century Athens, and late nineteenth-century Vienna. This thing, whatever it is, lurches unsteadily along, but it keeps moving, carrying us along with it, even at this turning of the transepochal in which, perhaps, the oedipal is being irremediably transfigured.

Psychoanalysis, like other institutions, must be generalizable if there is to be a transmission of technique, knowledge, and teaching, but it must again become radically particularized, in each session with each analysand, if it is to be efficacious. This is an impossible ideal, perhaps, but it raises the necessary question of the individual, or, more specifically, what Heidegger calls *Jemeinigkeit*: the own-most that is always related to futurity, the call of conscience, and death. "Because *Dasein* has *in each case mineness*," Heidegger explains, "one must always use a personal pronoun when one addresses it: 'I am', 'you are'" (1962, 68). Oedipus, too, requires the play of pronouns. *The* oedipal complex must be permutated into *this* oedipal complex; and, in a sense, Oedipus, which is Freud's name for every human being, must also learn to take on the name of each of those beings. Oedipus, the passage between the universal and the particular as idealized limit-concepts, has the structure of language and the subject in-formed by language as the joining moment of *langue* and *parole*. An utterance available from the history of the world, it must also be time and again idiomatized.

Such transfers can only occur because there is a structure of comparability between cultures, languages, epochs, individualities, concepts, and so on that has already been, and must further be, opened up as a space for the transference of meanings. There is, and has always been, a web at work that is a necessary prelude to the play called *Oedipus Tyrannos*. Freud, for instance, notes that the "action of the play (*des Stückes*) consists in nothing other than the process of revealing, with cunning delays and ever-mounting (*schrittweise*, step-by-step) excitement—a process that can be likened (*vergleichbar*) to the work of a psychoanalysis—that Oedipus himself is the murderer of Laius, but further that he is the son of the murdered man and of Jocasta" (1969, 295). Cunning delays; mounting excitement; then the release: such is the play, such is the Freud-Fliess affair, such is psychoanalysis. Analysis—indeed, subjectivity itself—is unthinkable without this step-by-step, without the temporal delays at work in signification.

Such equivalencies—Freud is forging a rather long chain of metaphors that will bind the Oedipus into a unity—can occur only within an already opened space of exchanges that enable comparison to occur. Freud is not, he claims, a philosopher, but he does offer in the letter of October 16, 1895, in which he is "almost sure [he has] solved the riddle" (exactly two years before the more famous letter), a "few pages of philosophical stammering" (1954, 128). This is a vocal analogue to the *flickering* of the light of consciousness, about which Derrida, commenting on the Mystic Writing Pad, has noted that "The becoming-visible which alternates with the disappearance of what is written would be the flickering-up (*Aufleuchten*) and passing-away (*Vergehen*) of consciousness in the process of perception" (1978, 225). This stammering-flickering of voice and eye characterizes Oedipus and all those who follow in his erratic footsteps. Speaking without interruptions is impossible.

Psychoanalysis, philosophy, and literature—and which is Oedipus?—must all account for, and proceed with the help of, this stammering, with terms such as Heidegger's *die Fuge* ('jointure')[1] and Derrida's *différance* that gesture toward the (non)fact that there must, for a (meta)psychology to occur, be a clearing of Being, the ontological difference between Being-beings, the spacing-temporality of writing, and a network of potential linkages that allow likenesses to be drawn from the nanomachinery of differences produced by, and constituting, the soulful apparatus of the psyche. But as such comparisons occur—between, for example, an old play and a new healing technique—these, in turn, by turning the lathes in the workshop of language, open up a more encompassing space for the shaping of meanings. Such is Freud's poeticizing of an Oedipus that has traveled prodigious distances back and forth in time (as if time were a line called *fort* at one end and *da* at the other end of the terminus, but that's a joke).

Within this space of disclosed translatability, Freud discusses the *Oedipus* in the *Traumdeutung*. What is it? We know, we think, that it refers to a play by Sophocles (a text? a performance? a fragment?) and that this play is transmittable across a multiply fissured and connected history. Freud makes a link between the play and his own work, noting that his discovery that all children, during one stage of development, love one parent and hate the other, is

> confirmed by a legend that has come down to us from classical antiquity: a legend (*Sagenstoff*) whose profound and universal power to move can only be understood if the hypothesis I have put forward in regard to the psychology of children has an equally universal validity. What I have in mind is the legend (*Sage*) of King Oedipus and Sophocles' drama (*Drama*) which bears his name (*das gleichnamige*). (1969, 294)

The play, then, "bears the name" of the legend. It carries the legend, which has no point of discernible origin, onward to do its legendary work.

Two narratives in a kind of diploidic structure share the same name and begin to multiply. (Oedipus is always a multiplier and a divider.) The story has just begun and there are already "legend" and "play" within the larger comparison between Freud's "richly counted" experience with children and the legend from antiquity. Freud is constructing a series of intersecting analogies as he, in a logocentric manner, constructs the Oedipus. Psychoanalysis, emerging as Freud writes, is being compared with history, literature, philosophy, politics, and religion as it draws all of these discourses into its own gravitational field.

Freud's lexicon makes use of the oral legend (*Sagenstoff, Sage*), the Sophoclean drama (*Drama*), and, all in quick succession, the tragedy (*Tragödie*), the tragedy of destiny (*Schicksaltragödie*), and the fable (*Fabel*). In footnotes added in 1914 and 1919, the Oedipus *myth* and the Oedipus *complex* appear as analysis generates a more technical, regulated, and standardized discourse. And there is one, and only one, usage of the term *Trauerspiel*, literally the "play of mourning" (but translated as "tragedy"). As Walter Benjamin says of the original seventeenth-century German plays: "[T]hese are not so much plays which cause mourning, as plays through which mournfulness finds satisfaction: plays for the mournful" (1977, 119). And, in agreement with Freud, he asserts that

> Nothing is in fact more questionable than the competence of the unguided feelings of "modern men," especially where the judgment of tragedy is concerned . . . the modern theatre has nothing to show which remotely resembles the tragedy of the Greeks. In denying this actual state of affairs such doctrines of the tragic betray the presumption that it must still be possible to write tragedies. (101)

Freud agrees with Benjamin's estimate of the impossibility of modern tragedy, so he turns back to the Oedipus for the authentic exemplar of tragedy.

Benjamin says of the *Trauerspiel*, "Historical life, as it was conceived at that time, is its content, its true object. In this it is different from tragedy. For the object of the latter is not history, but myth, and the tragic stature of the *dramatis personae* does not derive from rank—the absolute monarchy—but from the prehistoric epoch of their existence—the past age of heroes" (62). It is in the region of this slippage—and it is a necessary slippage—between history and myth that Freud works to construct both the history and the myth of psychoanalysis. *Oedipus* is, like the *Trauerspiel*, a pattern of mourning, a name for the rite of initiation into enculturated individuality that always entails a difficult, more or less successful passage through the sorrow of finitude. (That there may be certain types of joy within finitude must wait for the journey to Colonus.)

Freud uses the terms for different genres from different historical periods—*Sage, Drama, Tragödie, Schicksaltragödie, Trauerspiel, Fabel*—not just as a conscientious writer's habit of substituting one word for another for the sake of style, but also, and much more importantly, as a way of constructing the universal primordiality of the Oedipus. In a quite Hegelian manner, Freud gathers a variety of genres, and their multiple historical differences. He employs diverse genres and under the name of Oedipus constructs the signified of the "tragedy" that from the psychoanalytic perspective will signify the necessary symbolic work faced by every proto-individual. And it is, perhaps, the very possibility of such work that cybernetics threatens.

Freud's treatment of the Oedipus does not, of course, remain a critique of the Sophoclean text; his literary criticism serves the psychoanalytic project, and he pushes with a certain urgency to ground both the literary forms and the oral legend on the more primordial experience of the fantasy of loving the mother and killing the father. There are, he notes, the "primeval wishes of childhood (*urzeitliche Kindheitswunsch*)" (1969, 296), the "primeval dream-material (*uralten Traumstoff*)" (297), and "the reaction of the imagination (*Reaktion der Phantasie*) to these two typical dreams" (298), all of which give rise to a primordial template for the Oedipus legend and its subsequent formalization by Sophocles.

Phantasie itself is fractured and fertile, marking one of the limits in Freud's text of his romanticism. Speaking of romanticism's interpretation of symbol and myth, Derrida has written that to treat metaphor as

> the trope of resemblance . . . is above all to take an interest in the nonsyntactic, nonsystematic pole of language, that is, to take an interest in semantic "depth," in the magnetic attraction of the similar, rather than in positional combinations, which we may call "metonymic" in the sense defined by Jakobson, who indeed emphasizes the affinity between the predominance of the metaphorical . . . and romanticism. (1982a, 215)

This is *one* of the limits of Freud's text, for as he passes through the dream book he very much intertwines the likeness difference of both metaphor and metonymy.

The *ur-*, for its part, takes this experience to the very edge of time, perhaps to the placed-moment when time comes into being as the precondition for the unfurling narrative of identity that is organized around parental hatred and the desiring aggression of the child. And this *ur-*, in Freud, is related to the *nach* of *nachträglich*, which gives "directionality, sequence, and mode" to the spaced temporality of experience. The former can only appear in the guise of the latter, and without the latter we can have no perception of the former. There cannot be the edge of the not-time (primordiality) without the temporalization of narrative. This positions the Oedipus at the shimmering boundary between nature, which does not know time, and culture, which opens up within temporality.

Nature becomes culture, in this passage, through the bridge of the wish. "Like Oedipus," Freud writes, "we live in ignorance of these wishes, repugnant to morality, which have been forced upon us by Nature, and after their revelation we may all of us seek to close our eyes to the scenes of our childhood" (1969, 297).[2] This may also be part of the wish of technology in its desire to cut itself off from the "primitive" in all its guises, to create a subject without a childhood with its terrors and dreams. And this as the eyes, once blind, are beginning to open through the spectacles of Freud's writing. Nature forces libidinal wishes upon us that cast a veil over our sight, a veil that will require a revelation to be torn away. Oedipal logic (the logic of metaphysics) governs the rhetorical unveiling, which is revelatory violence, of the Oedipus. (There is a great deal to be said about the desire to [un]veil—think of the trajectory from Schiller's and Novalis's images of the goddess at Sais to Heidegger's *aletheia*— but that will have to wait.)

This logic-rhetoric exhibits a reflective self-sameness. Oedipus is the self-enclosed, with the only exit from the circuitry of death being to enact the violence that had been established long before birth in a repetitive ravaging of the eye with a filial prick of the mother's brooch and a long, almost interminable, exile. Until the rumors of a disappearance that, they say, is neither nature nor culture. Something other, we have been told, but something that can't be said. Gift, he said, ahead of time. Gift.

Let me slow down and backtrack. In Freud's construction of the Oedipus in the *Traumdeutung*, there is a clear causal-temporal hierarchy in place: nature, the complementary infantile dream wish about mother and father, the imagination that (somehow) shapes the wishes into narrative, the oral legend, the Sophoclean text performed for the lucky Athenians, and then the Sophoclean text that— through uncountable byways and however much it has been corrupted—has passed into the German that Freud recognizes *as a match* with his own experience with psychoanalysis. It is, to say the least, a multiply mediated match that moves from wish fear to imaginative structure to orality to writing to performance to reading and then back down the line, in that backwards flow, that twisting of subjectivity, without which analysis is unimaginable. It is no accident that the Oedipus becomes the kernel of psychoanalysis, for here, as compressed as nuclear material—that awaits the work of fusion and fission—we see both the logic of the dreamwork as well as the logic of analysis itself that traverses the dreamwork. *Both* of these logics carry, and are carried by, the name of Oedipus.

Freud himself is snared in this network of double identifications, with that of the child who loves one parent and hates the other *and* with the Oedipus who learns, through intellectual work and suffering—"Every line I write is torture" (1954, 210)—to read his fate aright and thereby release himself toward that fate, as inexplicable as it might be. In this section of the *Interpretation*, Freud is constructing a rational argument, and the discovery of the love-hate relationship toward the parent is, as he says, "confirmed" by the fabulous play. There is an "unmistakable indication" within the text itself that is the "key to the tragedy and a complement to the dream of the dreamer's father being dead" (1969, 297). Freud, developing an argument within the standard procedures of reason, uses the legend and the play (forms of secondary revision) to prove his thesis about the non-rational primordial wishes that spring from the edge of nature.

Childhood experience is first and the oedipal revisions are second, but both are mediated by Freud's experienced reading of patients and plays. The first, however, is merely a fantasy among different possible fantasies without the corroboration of the secondary elaborations of art, which show "unmistakable" evidence of the first. Derrida has identified this (non)logic as the "logic of the supplement" that disrupts the usual order of events—such as "primary" and "secondary"—as they are arranged within the oedipal-dialectical enclosure of metaphysics. Freud marks one of the limits of the project of metaphysics, yet, necessarily, he continues to work within metaphysics even as, along with his predecessors Copernicus and Darwin (and the one he resisted, Nietzsche), he abrades the tradition with every arduous step. In these few pages of the *Traumdeutung* that organize the contours of psychoanalysis, metaphysics and the logic of supplementarity actively crisscross one another. Freud knows *Oedipus*, the play, before he is able—through his self-analysis, his experience with patients, and his friendship with Fliess—to name the structure of fantasy "oedipal." As a general structure of signification, the name Oedipus was, as it were, waiting for Freud when, in 1897, he explicitly appropriated and thus transfigured it.

And, at least by 1899, he recognized this preparation: "There must be something which makes a voice within us (*Es muß eine Stimme in unserem Innern geben*) ready to recognize the compelling force of destiny in *Oedipus*, while we can dismiss as merely arbitrary (*willkürliche*) such dispositions as are laid down in [Grillparzer's] *Die Ahnfrau* or other modern tragedies of destiny" (296). *Die Stimme des Gewissens* is the "voice of conscience, the still small voice," a voice that, however small, nonetheless has epochal resonances. Its origins are in 1 Kings 19:12, and Freud uses the phrase at least twice in his letters to Fliess. First, on October 10, 1895, he writes: "But the mechanical explanation is not coming off, and I am inclined to listen to the still, small voice which tells me that my explanation will not do" (1954, 126). And then again at the turn of the year: "A still, small voice has warned me again to postpone the description of hysteria—it contains too much uncertainty" (1954, 141). It is a voice of quiet warning, a Socratic *daimonion* or an oedipal oracle. A voice from outside the immediate stretch of rational analysis that, as it were, guides that analysis, keeps it on track through a meticulous caution. The implications of this sentence are endless—for instance, what is this "it" that "gives" the voice and its ability to recognize?—but let me focus on just the most pertinent for the question under consideration: what is the Oedipus?

There is a fit, Freud announces, between this inner voice and the destiny of Oedipus; the two are conjoined in a contemporaneity that requires both a con- and contre- of *temps*. It is at this point in the argument that Freud asserts that the "inner voice" recognizes the nonarbitrary, primitive, fitting, fantastical, truthful impact of the material that moves moderns as much as it moved "contemporary Greeks." There is a profound jointure that he has unveiled between the ancient world, (post)modernity, and each human being. This jointure is neither the universality of sameness nor the absolute difference of nonrelationality, for time is neither a circle nor a line. We are all, in this sense, enjoined as contemporaries.

On the other hand, he asserts, there is a lack of fit, no pronounced emotional response, to modern tragedies like Franz Grillparzer's *The Ancestress* (1817).[3] Freud, perhaps, is too obsessed with tracking down his ancestor to allow a response to the dark continental mass of the ancestress, and certainly Grillparzer's art is not up to the sophistication of Sophocles'. (Who is Freud's ancestor, Sophocles or Oedipus? Does it matter? What is the relation between the two, the so-called historical and the so-called fictive?) In any case, Oedipus is "compelling"; Grillparzer is "arbitrary." Oedipus is a relic of antiquity (*das Altertum*); Grillparzer is, among many other writers, modern, and Freud does not collect the art of any avant-garde. But this is not Freud's only comment on the old quarrel between the ancients and the moderns.

Before mention of *die Ahnfrau* (and recall this is part of what he already said in his letter to Fliess), he had already argued that "Modern

dramatists have accordingly tried to achieve a similar tragic effect by
weaving the same contrast [between divine will and human power-
lessness] into a plot invented by themselves (*selbsterfundenen*). But the
spectators have looked on unmoved while a curse or an oracle was fulfilled in spite of all the efforts of some innocent man: later tragedies of
destiny have failed in their effect" (1969, 295). Oedipus "deeply moves"
the spectators; modern drama, created from a misunderstanding of
the Sophoclean conflict, leaves spectators "unmoved." Modern drama
is self-founded, and therefore arbitrary and ineffective. Oedipus, on
the other hand, is *not* self-founded, but given, at least in its rudimen-
tary form as wish, by nature. Oedipus, for Freud, encounters his fate
beyond the "mere" subjectivity of self-founding as he shatters against
the terrible gift of nature and the voice of Apollo.

This is a very strange moment. As Freud writes the *Traumdeutung*, he is laying down a significant track along which (post)moder-
nity will gain momentum. In this function of a steel-driving man, he
is employing reason to illumine the dark regions of the psychic machine and of human conduct, thereby bringing the instincts under the
sway of rationality: "*Wo . . . es . . . war*," I shall be. This is the project of
the *Aufklärung* that opens the path, through science, toward the ma-
turity of humanity. And yet for Freud the project of modernity—
which, as Goux has argued, might be said to have begun with the
oedipal perversion of the monomyth of ritual initiation—goes astray
in its willful arbitrariness insofar as it claims to be "self-founded."
There is a deeper set of tracks, for Freud, laid down by nature than
the modernity of the self-transparent cogito, and yet the cogito, rep-
resenting all modern forms of subjectivity including the psychoanalytic ego (although *not* the psychic apparatus as a whole) must, as it
were, set up shop where the it-was was (and is, although differently).

Freud's psychoanalytic work that founds (post)modernity also,
at the same time, rejects modernity in favor of a version of classical
antiquity transported across untold cultural boundaries. The issue
at stake, in his interpretive positioning of the Oedipus, is neither the
modernist concern with autonomy nor the aesthetic formalisms that
create an effective drama, but, rather, the transliterations of libidinal
aggression and desire. Autonomy, for Freud, can only be an *achievement* of psychological work, a secondary revision that works over
the earliest formations of desire that are laid down before the cogito
is formulated, before there is even the symbolic opening of language
for the protosubject. The preoedipal is oedipalized.

Writing about one of his own neurotic episodes in June 1897,
Freud tells Fliess that it consisted of "odd states of mind not intelligi-

ble to consciousness—cloudy thoughts and veiled doubts, with barely here and there a ray of light" (1954, 211). The rational "rays of light" are certainly what Freud projects as he moves through darkness, and thus he is a modern. And yet as long as the psychic apparatus generates that flickering called human consciousness, the clouds and veils cannot be completely obliterated. Rationality is bounded; it works within a given context, but it is not the whole. Freud is snared in a contretemps of time that shatters any presumed unity of experience *and* allows—in ways I do not fully understand—a cobbling, stitching, and caulking together of discontinuous temporalities. Like the rest of us, Freud is, simultaneously, the primitive, the romantic, and the (post)modern. His work on the Oedipus, which is akin to his work on the *tyrannos* itself, is like the creation of a *Sammelperson*, a composite figure that requires de-composition through an interpretative listening that hovers as evenly as it can.

Oedipus is a name, for Freud and his legatees, that binds together differences. He conflates a variety of genres: legend, classical tragedy, *Trauerspiel*, and dramas of fate. He combines literary forms with his own clinical experience, children's fantasies with adult neuroses. And, in perhaps his most audacious move, he discursively transforms vastly different historical epochs into that one universalizing epoch called "oedipal." Recasting the meaning of history, myth, civilization, subjectivity, and temporality, Freud sets out in the *Traumdeutung* to cross the Acheron, the flowing boundary of Hades. And what he discovers, this *caulking* along the seams of his craft, is the compelling voice that has long been calling to his own deepest inner ear, projecting forward the meaning of Oedipus.

The temporality of psychoanalysis, so often equated with an archaeological dig into the recesses of the past, is, in its fundamental directionality, futural. As Weber remarks about Lacan:

> The temporal structure of the subject that Lacan's reading of Freud strives to articulate, stands in marked contrast to this perfected present [of Hegel]. The perfect tense is supplanted by the future anterior, thus calling into question the very foundations of subjective identity conceived in terms of an interiorizing memory. In invoking the future anterior tense, Lacan troubles the perfected closure of the always-already-having-been by inscribing it in the inconclusive futurity of what will-always-already-have been, a "time" which can never be entirely remembered, since it will never have fully taken place. It is an

irreducible remainder or remnant that will continually prevent
the subject from ever becoming entirely self-identical. (1991)

This voice from beyond that says "Oedipus," and which is stamped on
Freud (he claims) by nature itself, is the nonfigurable, always figured
nexus of the net of existence. It whispers to him a tale of an oracular
voice, a plagued city, abandonment, rage, murder, incest, the ferocity of
the rational, and exile. It grants sight and guarantees blindness.

   And Freud himself? He knows that nature requires the psychic
machinery of writing, and, therefore, he "performs for us the scene
of writing. Like all those who write. And like all who know how to
write, he lets the scene duplicate, repeat, and betray itself within the
scene" (Derrida 1978, 229). Furrowing his brow on an October
evening in 1897—already past forty and with his face a network of
wrinkles—he bends over his paper with pen in hand and calls, for
the first time, this having-been called the Oedipus.

Oedipus and Technological Surveillance

   "Oedipus" is used in more ways than we can count; he (or it) is
uncountable, as well as unaccountability productive. Oedipus is a
multiply divided and repetitive enigma that prompts an incessant
murmuring through as many voices as can be called upon to utter
the name, a name that surely did not come from the mother or the
father, but, perhaps, from outside the fold or from the other family.
Oedipus is living and dead—and neither living nor dead—existing
only along that boundary line at which language positions us, sub-
jecting us to the constrained freedom of speaking subjects. Oedipus
is the subject of his own, which means our own, discourse about our
place in the world. And he is also, like the rest of us, subject to the
discourse of many others, including his own forgotten history. As
are we all.

   Oedipus, to repeat, is uncountable. Technology, on the other
hand, is what really counts for us. It makes things, including
human things, count, by counting. It makes sure we are tagged and
numbered by placing us in the database. It creates value by counting at ever faster rates of speed, by digitizing, accumulating, archiv-
ing. By buying and selling that about us that can fit on a disk and
travel the wires, echo and ricochet from satellites orbiting our point
of arrival and departure. Oedipus sets the scene for technological
surveillance. He, or it, accomplishes this, in part, by installing a

self-surveillance and self-guidance system called rationality within both his own interior and, by becoming the police, within that of the polis. Oedipus moves, without at first recognizing it, from inside to outside the polis, back inside and then outside once again, cast into exile by his own word. Fleeing from the oracular voice of Apollo, who spoke enigmatically long before Oedipus saw the light of day, the man of ideas passes through the riddle of the half-woman, half-beast and enters into the enraged turbodrive of the rational, of the "one who will know."

He is willing, we must remember, to torture the old and the blind, whether shepherd or bird-reader, in order to get what he wants. He will track down the truth, regardless of the costs. He will not count the costs; he is beyond that type of reckoning. He keeps a lookout for traces, clues, in order to turn them into his own autobiography, and to learn at last, through the logic of *Nachträglichkeit*, the truth of what he has lived. Oedipus exhibits a relentless tenacity. He bares his teeth and snaps closed the trap of his mind, not letting go until it (*das Es*) has had its fill, after which it might give something forth—an enigmatic gift, for instance—out of the mystery of his passing.

And that is noble, admirable. Oedipus is the passionately driven, the compelled man of insight, the man who forges a discourse on a method that demonstrably works and thereby heals, if only temporarily, the polis. These are things we all know and therefore must continue to repeat, but within the repeatable, as we all also know, there lies both the necessity of erotic thinking-writing as well as the rending voice (if it is a voice and not simply a mute force) of death. Oedipus's consciousness, his unconsciousness, and his epistemological rage for order all bring with them death. When we speak the name of Oedipus, we evoke the *tyrannos* and enter, without choice, into an interminable revision of the familial-political narrative that provides the conditions for self and other-knowledge, a violent search that will never move beyond, but always stay circulating within, the political-familial, the *oikos* of the economy of desire and death. This is the locus, too, of technological surveillance and its logic, for the Eye has become multiply instantiated as a video grid that, like the law, is always turned on. Oedipus, as the representative of what Goux has named the "prototypical figure of the philosopher" (1993, 3), represents the "unidimensionality of the Apollonian passion" (106) and is already the site of what will be instantiated as technosurveillance. Reason, in this form at least, carries with itself the promise of lookouts and listening posts.

Power is on the lookout for its rivals, its transgressors. The video cameras peer down like dark cyclopean eyes, gathered together at command central, to cover every angle of movement. The dream of the surveyors is to survey every scene without remainder, especially (but not only) those scenes where the outlaws might act (assuming the outlaws are not the in-laws). At least that is the claim. But the transgressors always look back at the law (and neither term is a simple concept; each must be articulated into its many forces). Each rival is always looking ahead, trying to outmanuever the other in the realm of control, predictability, and technical imaging, and, inversely, working incessantly to find ways to act without being seen by the other.

As Bar-Ami Bar On says of the strife between terrorism and its counterpart, "These various practices instituted by states to combat terrorism confine and isolate the citizenry and put it under constant surveillance. Surveillance makes one feel simultaneously safer and more vulnerable. . . . The effect is to induce in those who are observed a state of conscious and permanent visibility that assures the automatic functioning of power" (1991, 486). "Permanent visibility," in other words, is not only reserved for terrorists, but for all the passersby in the everyday life of city streets and fast-food joints.

The wished-for technocracy of the visible creates a concomitant technocracy of the invisible. The two desires—to see the other absolutely but not to be seen—act within the bounds of the network; both "visibility" and "invisibility" have meaning only by reference to the technological expansion of the senses, the prostheticizing of eye, ear, nose, and hand. *Visibility* no longer refers either to the simple empiricism of experience here and now beneath the sun, nor, of course, to that invisible sun of Platonism that renders the world itself, as world, brightened and intelligible. Visibility, as the foundation and goal of technosurveillance, requires, among other things, a *different temporality*, a different relationship to time than either the evanescence of experience, repeatable only through the difference of memory's creative transfigurations, or the "eternal idea" of the *archai* of the forms that indicates an "outside" or a "beyond" of time. Visibility is recorded in hours, minutes, and seconds. We must be able to temporally locate what is happening in the space of the camera (a term I will use as a metonymy for all the instruments of surveillance).

Click.

The "society of the spectacle" assumes it knows the answer, or at least the correct methodological response, to Nietzsche's question: "What if truth were a woman—what then?" and it replies, with a certain vulgarity, "We'll catch her in the act!" The machinery of love—the play between Eros and Thanatos—entails unveiling the veils so that the act itself (any time, day or night) will be visible on the screen (which is itself a word that slides throughout the projection of the spectacle). Technosurveillance always occurs within the familial-political, as the serial drama of murder, romance, and punishment, complete with the seduction of tripods and cameras. Private dicks are on the prowl, paid for their time and looking for dirt. The lens cap is removed and the button clicks; the closed circuit camera silently rolls, and then there's the fetish of footage. An image is frozen in time, becomes a commodity for sale.

Oedipus *is* the age of technological domination.

The Soulful Machine of the Psyche

As we have seen, Oedipus as a figure of what will becomes psychoanalysis comes to us first through Freud's pen in a letter to Wilhelm Fliess on October 15, 1897, and now, because Freud has already named him as his most entangled knot and tied it into a cultural loop, we are able to name Oedipus once again in a denomination both familiar and strange. On March 15, 1898, he writes to his "Dear Wilhelm" of his blind "burrowing in a dark tunnel," saying to his friend that "At this stage I am just plain stupid. . . . First I must read more about the Oedipus legend—I do not know what yet" (1954, 248, letter 85). How familiar that stupidity feels, no matter how long we have been hearing about, and listening to, the voice of the Theban-Corinthian beggar king. Freud goes on to articulate "Oedipus" as a sign of desire, power, and identity—of the phallus and castration—and Lacan, rereading, gives Oedipus another appellation in which we always hear an appeal: the Name of the Father that reigns over, and within, the Symbolic, bringing with it the plague.

Freud, like Oedipus the detective philosopher, creates a *method of inquiry* that limps back and forth, through ravines of silence, toward some theory that resembles a cure. All methods involve techniques, and psychoanalysis is a technique, a complex machinery. If the way is a technique, then the subject / object of that way must have a technical affinity with the method. It does: Freud has named the psyche a "soulful apparatus" (*ein Seelische Apparat*). *Seelische* can mean psychic, mental, spiritual, or emotional, but I prefer, as a provisional rendering, "soulful" so that the full force of the uncanny linkage between the "soulful" and the "apparatus" can be emphasized. Once again, as we saw in the discussion of the *Grundrisse* and Bartleby, the machine itself, as the symbiotically hybrid operation of the (non)human, is soulful (although this does not, as we will see, rid us of the problem of exploitation).

From its very inception, the apparatus of psychoanalysis is a technologics, wired for sound, making use of transatlantic cables, and bugged from the start. Because it is bugged, because microphones have been inserted everywhere, into every orifice and along every surface, psychoanalysis has begun to learn—and this beginning may never give way to what is known as knowledge— how to debug the rooms, the hallways, the frames and mirrors, the artworks, the attics and the acropolises, the ruins and the back alleyways. Or, if not how to free an area from bugs, if not to make a clean sweep by sweeping for mikes, then at least to learn, bit by bit, about the functioning of the listening / speaking apparatus, the switchboard of the soul, in both its hidden and its unconcealed aspects. To whom, or to what, does this work—these workers who are all of us—report? Is there a Central Intelligence Agency that sorts the information picked up from the listening posts, individual and collective, and then decides what it means, how the data is to be used? Is that the contemporary form of the hyperobsessive technological *Über-ich*?

Psychoanalysis is enmeshed and implicated in the world of secrets, power, surveillance, and listening posts that are installed, silent but recording, just behind the couch. But perhaps there is an essential, if only just the slightest, difference between the two, enough difference to make it count. Analysis is a technical apparatus that reads the drive of desire through an active ear and then releases the subject, or assists the subject to let go of its own accord, toward a certain type of rehistoricized freedom. It does not desire to store the archive indefinitely in order to be used against, or to simply track, the subject, but rather to recompose the archive of memories in order to enable a larger space for the play of the subject. It may not provide enough of a difference to make a difference, but, then again, it might. Analysis, one hopes, can provide a different sort of listening post and speaking apparatus than that solely governed by the technocapitalism of the world system.

Ein seelische Apparat, then, Freud says, wired and preprogrammed by the others. It generates energy through the psychoanalytic corpus that becomes visible throughout his writing career. "On Aphasia," published in 1891, concludes with the following sentence:

"It appears to us, however, that the significance of the factor of localization for aphasia has been overrated, and that we should be well advised once again to concern ourselves with the functional states of the apparatus of speech" (105). This concern about language and an apparatus, which will form Freud as a psychologist rather than as a neurologist, leads through the "Project for a Scientific Psychology" (1895), which perhaps most fully reflects Freud's fantasy of the machining of the psyche and which was begun, not accidentally, while returning home on the train after a meeting with Fliess. These rail links, these pathways laid down in advance that shuttle him to and fro, usually to meet with brilliant men, will appear and reappear in Freud. And there will be, finally, a train that leaves Vienna for good.

On September 23rd the Viennese writes, in a tone that can almost be called love, to the man from Berlin: "The only reason I write to you so little is that I am writing so much for you" (1954, 123). The "Project" is to be a gift, and in early October the "clinical knowledge of [repression] has incidentally made great strides," even though the "mechanical explanation is not coming off" (126). Freud is beginning to doubt his proposal and lays "the thing" aside again in the middle of the month. Then, on the twentieth, "the barriers suddenly lifted, the veils dropped, and it was possible to see from the details of neurosis all the way to the very conditioning of consciousness. Everything fell into place, the cogs meshed, the thing really seemed to be a machine which *in a moment would run of itself*" (129; italics mine).

The beatific vision of the psychic apparatus had occurred, and Freud sees, or thinks he sees, the thing behind the veils. It veritably purrs, as Freud, humming with delight, quotes poetry:

> *Was man nicht erfliegen kann,*
> *muss man erhinken . . .*
> *Die Schrift sagt, es ist keine Schande zu hinken.*

Strachey translates and explicates: "'What we cannot reach flying we must reach limping. . . . The Book tells us it is not sin to limp.' From Rückert's *Makamen des Hariri*, also quoted at the end of *Beyond the Pleasure Principle*" (1954, 130). Flying and limping; a book and the sinlessness of the limp. Oedipus, a man who flies, limps, and produces books, is certainly nearby, and Freud generously ends this letter with the promise to call his next child, if it is a son, Wilhelm. "If," on the other hand, "*he* turns out to be a daughter, *she* will be called Anna" (Freud 1954, 130). Oedipus, the confuser of pronouns. What did Anna think, if she did, of her shadow other, the Wilhelm that accompanied her throughout life, but never saw the light of day?

November rolls around, and Freud is making discoveries about the etiology of hysteria and obsessional neuroses, but, still, the

> time has not yet come to enjoy the climax and then sit back and relax. The later acts of the tragedy will still demand a lot of work from

Your

Sigm.

Who sends cordial greetings.

Mid-November, a bleak and blustery month of rain with the temperature falling toward the zero point, brings disillusionment about the hopes for the psychic machine. "Since putting the psychology aside I have felt depressed and disillusioned—I feel I have no right to your congratulations. Now I feel that something is missing" (134).

The thing, it seems, has become deflated; the vision of the cogs whirring behind the veils has sputtered. But the end of the month, bringing with it a break in the weather, finds Freud "in top working form . . . beautiful things, all sorts of new material." The apparatus now seems to him "a kind of aberration" and Freud has "become acutely aware of the distance between Vienna and Berlin." The trains are not running, except to carry the post as a substitute for the journey itself, but even "writing is so difficult." Nonetheless, Freud looks forward to "hear good news of you, wife, child, and sexuality through the nose" (135).

At the turn of the year, Freud offers a revision of the "Project," linking it directly with Fliess's ideas about the "state of stimulation in the nasal organ" (144). He includes, as a New Year's gift, Draft K, which concludes with the statement that "it is a question in the first instance of there being a *gap in the psyche*." And then, after a month's lapse, he opens his next letter, of February 6, with the fact that "There has been an unconscionable break in our correspondence." Since no one else is blowing it, Freud begins "blowing his own trumpet" and starts writing Fliess "immediately" (155). The sound of horns, all those upright and polished instruments of gold, is inspiring to him.

From this early period, then, right up until the end of his life, Freud works to clarify the organizational dynamics of the psychic mechanism, recognizing that the mechanism itself, the localization theories in the aphasia book and its many analogies, does not provide a sufficient explanation of the logos of the psyche as he envisions it. As he notes under the category of the "Psychical Apparatus" in the opening pages of *An Outline of Psychoanalysis* (1938): "We know two kinds of things about what we call our psyche (or mental life): firstly, its bodily organ and scene of action, the brain (or nervous system) and, on the other hand, our acts of consciousness, which are immediate data and cannot be further explained" (1974b, 144). There is the scene of action, and then there is another stage, that other *Schauplatz* that is quite inexplicable to certain types of the calculation of cause and effect. There is, paradoxically, something incalculable about the very calculability of the psyche. Freud, the heir of both rationalism and romanticism, sets this in motion with the *un-*.

The road signs at the oedipal crossroads point in every direction: Vienna, Paris, London, Buenos Aires, New York, and to every beyond, since we are in the midst of an apparatus, or more exactly an interlocking set of machineries, that has long been globalized with the hope of universalization. Oedipus is on the move. Freud, however, is standing still for a moment, looking out at the vista and congratulating himself as he wipes his brow in preparation for a greater, more demanding effort. He stands at the most determining and determined spot of the road that has been constructed by the metaphysical tradition where journey, light, and the explication of explanation meet. He is at a crossroads and, like his predecessor, will pursue the truth wherever it leads. He will walk along the road of his talking-writing/listening-reading method and see what he encounters. And the method, as we know, will create its own effects, of which we—the you, I, they, and it—are one.

Freud knows, already, that the psychic apparatus is a relational network of forces that must be understood not atomistically—the old science and philosophy are dead, though still glowing among the embers—but only as what Derrida will come to call, when he responds to Freud's call (among others), a network of *différance*.

THE PROSTHETIC CULTURE OF TECHNOCAPITALISM

Culture, with all of its power, seems always to just limp along, showing its barbarous history, but not often, recognizing it. It always seems to need a magic staff, a white cane, a hearing trumpet, or a mechanical voice box to amplify the sound. The chorus in

Sophocles' play, for example, justifying themselves, chants through their masks that "A god was with you, so they say, and we believe it—you lifted up our lives" (*prostheke theou*), and Goux notes that "*Prostheke* means addition, supplement, something added, aid, assistance" (1993, 18). With his own prosthesis in place, Freud has famously explained that as man sought to attain various cultural ideals, envisioned as gods, he has become a "kind of prosthetic God. When he puts on all his auxiliary organs he is truly magnificent; but those organs have not grown onto him and they still give him much trouble at times" (1989, 44*)*.

Jacques Lacan, taking his turn, teaches that what we learn in the analytic experience is that the "genital drive is subjected to the circulation of the Oedipus complex, to the elementary and other structures of kinship. This is what is designated as the field of culture . . . " (1977, 189). Pleasure (and pain) organized; genitality Oedipalized, sliced and diced through the wound, the limp, the blindness of Oedipus; and, then, circulated like love or money: that's culture. That's the culture that requires, and has always required, a prosthesis: a god or technology. Something to lean on, to enlarge our domains and increase our value, to enhance the senses. This culture, of course, is governed, though not seamlessly, by certain forces:

> Perhaps the features that appear in our times so strikingly in the form of what are more or less correctly called the mass media, perhaps our very relation to the science that ever increasingly invades our field, perhaps all this is illuminated by the reference to those two objects, whose place I have indicated for you in a fundamental tetrad, namely, the voice—partly planetarized, even stratospherized, by our machinery—and the gaze, whose ever-encroaching character is no less suggestive for, by so many spectacles, so many phantasies, it is not so much our vision that is solicited, as our gaze that is aroused. But I will leave these features to one side and stress something else that seems to me quite essential. (274)

The voice and the gaze, both part-objects for Lacan, are "planetarized . . . by our machinery." Multimedia is everywhere, both inside and outside the subject. Technology is no longer—if it ever was—a phenomenon external to the subject and the body of history; rather, as necessary prosthesis, it has always coconstituted the subject and history.

That essential something Lacan mentions is the "drama of Nazism" that had attempted to derail Freud, and he says, "I would hold that no meaning given to history, based on Hegelian-Marxist premises, is capable of accounting for this resurgence—which only goes to show that the offering to obscure gods of an object of sacrifice is something to which few subjects can resist succumbing, as if under some monstrous spell" (275). Technology and the "obscure gods" accompany one another; the establishment and instantiation of technological reason, oedipal-Enlightenment reason we should say, vanquishes neither the gods nor their twins, the monsters. Not, at least, from the Symbolic, which swarms with phantasms.

The body politic and the individual body of each subject are divided bodies. They are alienated both from themselves and from others, but it is just this always-having-been-divided that "makes necessary what was first revealed by analytic experience [experiment], namely, that the ways of what one must do as man or as woman are entirely abandoned to the drama, to the scenario, which is placed in the field of the Others—which, strictly speaking, is the Oedipus complex. . . . [T]he subject depends on the signifier and the signifier is first of all in the field of the Other" (204–5). The "I," with its gaze and its voice, depends for its articulation on the field of the Other, on language, on culture, on the Name of the Father, the phallus, the absence of the mother and of origin. What you will. The Other is the prosthesis of the subject, something necessary to incorporate and carry, but also something that never fits perfectly and is therefore unbearable. It chafes. It scars. It is called, by some, Oedipus, but who names Oedipus? His mother can only call him the "man of agony." When things are bared, whether teeth or the body, there is the anguish of namelessness seeking a name, the indeterminate breaching into, if that's possible, the determination of the finite. Oedipus, counted out of the game by his parents as they attempted to cast out the voice of Apollo, must violently expropriate himself of his illusory history in order to reestablish the boundaries of the culture of the city.

Culture is the field of Oedipus, of delayed, waylaid, and misinterpreted relations of desire, aggression, ignorance, knowledge, oracles, and the imaginal. Of boundaries that are necessarily, but unwittingly, crossed. We, as individual subjects, are never quite individual; we, as individuals, are always more than individual. *Dasein* exists just as this "never quite" and "always more than," which shows itself as a crevasse that requires us to learn courage and a knack for leaping. We become individuals because we are mediated

subjects that exist alongside the others that we add on and subtract,
one of which is the technological apparatus of technocapitalism that
does not exist in some "outside" of subjectivity.

Technology is one side of the sideless Möbius strip of which
Lacan is so fond. As oedipal, modernity (which in this sense begins
with the pre-Socratics) forms reason into the instrumental (Freud
also speaks of the *Seeleninstrument*), and as Goux argues, "By constructing illusion through technical means, by calculating the rational
conditions of *trompe-l'oeil*, the Greeks disenchanted the world
through re-presentation itself" (1993, 131). Re-presentation, the dominant mode of western being-in-the-world, disenchants the world over
time (perhaps to get over time as we get over a loss, perhaps somehow
fighting over time so as to leap over time to deathlessness) and
thereby becomes the instrumentality that *desiccates* the world
through ratiocination and creates the conditions for the cybernetic
trompe- l'oeil of a globalized virtual reality. This reality, an electronic
and computerized specular panopticon, gazes at us from all directions and fixes us in place. No wonder we are paranoid; but it is only
the enigma of ourselves turned topsy-turvy, us looking at ourselves,
askance, akimbo, and from perhaps an unbridgeable distance.

There are, as usual, consequences and implications, several of
which Goux identifies when he remarks that:

> In Oedipus's view, and Hegel's, man can break, once and for
> all, irreversibly, with those inferior and polymorphous
> realms, in all their discordant, enraged, or voracious har-
> mony; man can install himself in the reason that is his
> proper nature, and there he can remain, self-sufficient. The
> victory of Hegel's Oedipus is a victory of unity (a single,
> unique reason in the unified mind of man) over the danger-
> ous multiplicity of the passions, which are not confronted
> and consumed (in a fiery, bloody trial), but denied by an in-
> tellectual and self-reflective decree. (156)

Oedipus, Goux continues, "typifies that critical mutation achieved
by the Greeks." He destroys the "cryptophoric mode of symboliz-
ing," which presumably leads to a proper initiation and the establishment of a new order, and establishes the conditions in which
the "subject discovers the world as an object (rather than a sign)
and situates himself as a [rational, autonomous, democratic] subject" (120–21). With Oedipus and the age of technical reason in
which we live, the cryptophoric almost vanishes, for displacing un-

reflective mythic knowledge of initiation Oedipus-Freud twists the old *epistemē* into encrypted cryptography, thereby changing the meaning of all codes.

Working through Technocapitalism

Although there are certainly possibilities of individuals "working through" the oedipal complex in more or less successful ways, there is no "dissolution" (*der Untergang*, Freud writes, echoing Zarathustra's call from the mountains) of the technocapitalism within which, for us, the oedipal scenario—with its split names and desires, with its watchful enforcers of the (out)law—unfolds and refolds itself. As Lacan has said, "There is in effect no other way of accounting for the term *durcharbeiten*, of the necessity of elaboration, except to conceive how the loop must be run through more than once" (1977, 274). Oedipus is a name for a symbolic-cultural feedback loop that, as a form of practice, runs us through our paces time and time again. We are the bio-machine, the psychic apparatus that produces culture; we are the commodities, the artificial intelligences, produced by culture. "Look," it says, "here I am again," but this "I" is a mirage, a shimmering effect, quite beautiful at times, of the "here," the "am," and the "again," that appears in a seemingly simple imperative sentence: "Look, here I am again." Look. Look, here. Here am I. I am. Again. Again, look. I am I am. Again. Here. Look, hear.

Oedipus is the name of the law, which is neither inside nor outside the law and is both inside and outside the law, that loops us back into the networks of language, capital, and social relations, and that knows that the only "beyond" of technocapitalism is a returning to Thebes, the place where the word first sounded, with a difference, one name of which may be Colonus or 19 Berggasse. This is the place of suffering and work, of disappearance, mystery, and a gift to the city. It is not *accomplished* only by the conscious intent of the character, whatever his name or epoch, called Oedipus. But it nonetheless requires his presence, his presents. The gift requires that the stranger return home and recognize the strangeness of the economy of the home. It, the it of presents, is accomplished through him, as it were: his name marks the spot of a memorable, if finally inexplicable, event. An event that launches the attempt to explain but also erects the wall against explanation, perhaps the very wall that the projectiles of cybernetics hope to breach and bring down.

Explanation—shall we call it philosophy, science, psycho-analysis?—is not enough to "dissolve" the oedipal complex, whether individually or, through the individual, culturally. In fact, Oedipus is irreducible and inexplicable. We could not live without it as it gestures toward the inscrutable at the edge of the technics of technique. How, then, might we respond to the crisis of the technologos in which we discover ourselves? There may be nothing we can do. Human beings may be on the verge of a disappearance that will be simply an end, a loss without the recompense of a gift left behind or any interlocutors arriving from the future to wonder once again, "Where am I?" Assuming, however, that we might hand down a little something to those who are to come after us, how might we compose our small bit of text, shape our bit of debris, to confront the intrusions of technosurveillance that looks out only for itself?

We can continue to think, to look back, to learn how to grope our way blindly along, fingers gently touching the ground. We can think Oedipus again; think the world of calculation and wonder about the incalculable. We can attempt to think the nothing in its many distinctions and how odd it is that the nothing might be distinctive. We can listen to the dreams, to the speech of the unconscious, and do the difficult symbolic-cultural work of dissipating the destructive fixities of oedipalized desire. We can think, again, the pathways of desire, fear, and violence, all of which shimmer like a lure or a threat at each and every crossroads. We can continually be vigilant about the Law of the Name and the No, just as that law is vigilant toward each of us.

We can continue to educate ourselves about the technological and its significance, the ways in which through it we alter the world and its subjectivities, but this education, like Freud's journey in the *Traumdeutung*, will inevitably be accompanied by the deepest darkness. We can make the gamble that democratic accountability can act to offset the burgeoning expansion of the tools of surveillance and control. We can attempt to determine whether it's possible to do ethical work *through* the technological, mediated by the technological.

We can, as Goux and others suggest, recall the Sphinx and the "feminine" from the repressed sedimentation of patriarchal culture (although great care is needed to speak this "feminine" without essentializing it into a preinscribed place within the Symbolic). We can explore other fundamental metaphors than vision and sight. We can honor and mourn the body as it is being destroyed by its construction within the biometric grid. All of this work, in a different version of the *Unter-gang*—this work in our different situations with our varying capabilities—releases us, however slightly, toward the future.

We must continuously work to see through, by means of and beyond, the screens of oedipality as it coalesces, its pixels shimmering, on the screen(s) of transepochal culture. There is the necessity for continuously *(re)reading*, attending to the stops and starts, the shunting asides and switches, of signification rather than the petrified "message" of the signified. Or, as Weber writes near the end of his return to Lacan's return: "Which is why no formalization can ever entirely replace reading. And why reading never takes place in general. Reading in this sense . . . means relearning how to be struck by the signifier. And by its stage. In the theater of the unconscious, one never gets over being stage-struck" (1991, 151). On stage, then, and knowing our lines and timing will be off the mark, we must nonetheless keep acting, working to learn the play by heart and then release it to the fateful disappearance at Colonus.

9

Temps: Time, Work, and the Delay

THE TIME OF TECHNICITY

We are all temps.

Riding the tides of time that ebb and flow, settle and swirl. Weathered and scored. Scoured. "There is a 'now' of the untimely; there is a singularity which is that of this disjunction of the present" (Derrida and Ferraris 2002, 12).

Here today, gone tomorrow. Split. Timed out.

Provisionally employed, tensely employed in the legerdemain of time's play. We temporize: negotiate and compromise, adjust ourselves to circumstances. We are tempted and we tempt, always stretched along beyond ourselves, intoxicated by life and death and unable, not for much longer, if ever we have been able, to tell the difference between the two.

For the last time, then, a few amusing etymologies of "temps": 1. time (Obs) 2. in legerdemain, the exact moment for executing a required movement as when the attention of the audience is distracted by some other act. 3. tense (Obs) 4. Temporary employee. *Temporize*: 1. to suit one's actions to the time or occasion; to conform to the circumstances. 2. a.) to give temporary compliance or agreement, evade immediate decision, so as to gain time or avoid argument b.) to parlay with someone to gain time. 3. to effect a compromise *temulence*: intoxication. *Tempest, temple, tempt*: fr. L., to stretch. This will lead to the auguries of the tides, but that's another itinerary.

Plato taught us to hold up our fingers to begin the network of relationships that will create the web that twists implacably around each of us, dragging us along, reconfiguring body, world, and identity. One, then two. Counting gets underway *as philosophy* that reproduces itself and its world in a frenzy that rages, often quite serenely, across millennia, and, before too long Being is the calculable and *Dasein* exists only as a form of the *Gestell*, as the setup whose task it is to measurably control the earth and the regions beyond the earth.

Of course, it all takes time, and time takes all. But what is time? What is the time of technologics? "[T]ime 'is' not," Heidegger reminds us, "but rather temporalizes itself . . . [and] every attempt to fit time into any sort of being-concept must necessarily falter" (1984, 204). Time, in other words, cannot be defined; metaphysics cannot clarify the concept, because time is not a concept. Nevertheless, metaphysics cannot be faulted for not making the attempt. In fact, from a certain perspective, metaphysics has done nothing else except elaborate a reading of temporality. We have seen the time line laid down from Aristotle through Heidegger and Derrida.[1] Dastur, for example, shows that "If Heidegger does maintain the classical analysis of time into a threefold structure, still, past, present, and future no longer designate a succession of nows on the 'line' of time but instead equiprimordial modalities of existence" (1998, xxx). Critiquing Derrida's seminal work on writing and temporality, Wood argues that

> The terms *"différance"* and "trace" seem to be used as part of a negative transcendental argument to deny the possibility of any concept of time dependent upon the idea of the present—that is, any concept of time at all. However . . . it is clear that Derrida cannot want it described in this way—as a (negative) transcendental argument. It would . . . make *"différance"* into a ground and thus condemn it to the status of a new metaphysical concept. Indeed, having identified the concept of time as such with the metaphysical tradition, it is . . . surprising, although gratifying, to see [Derrida] referring to "pluri-dimensionality" and "delinearized temporality.". . . (1989, 331)

Thus the aporias of even the most sophisticated attempts at the deconstruction of time. It seems as though, even with the most rigorous *rewriting* of the formats of temporality—as, for example, *Ereignis* or *différance*—the line and its others all continue to function as simultaneous analogues of an X.

We who are crossing the lines stumble along, faltering in our speech and our steps, giving time thought, just as time gives us time to think. We, even as the "we" is transfigured, crossed over and out, are (not) temps. "As they linger awhile, they tarry. They hang on. For they advance hesitantly through their while, in transition from arrival to departure. They hang on; they cling to themselves . . . it aims at everlasting continuance and no longer bothers about *dike*, the order of the while" (Heidegger 1975, 45). The epoch of metaphysics, which provides us with the dream and the means of continuation, beyond the duration of the organic, is a will to power. "Humanity has already," Heidegger observes, "begun to overwhelm the entire earth and its atmosphere, to arrogate to himself in forms of energy the concealed powers of nature, and to submit future history to the planning and ordering of a world government" (57).

In this transepochal moment, *Dasein* is *suspended* in the elemental fires of a technocapitalism that through its planetary alchemy, its turbocharged and furious rationalization of all horizons of individual and social life, radicalizes the (post)human dream in a manner that has, at last, begun to develop the specific instruments to accomplish its ancient desires. The wish for immortality, the death of death, that has in prior epochs been expressed by the projective doubling of the symbolic in the world of the perfect *eidos* or of the gods has now come to a new set of crossroads through which we are leaving the planet behind, leaving the organic body behind, and leaving behind the divisions between natural and artificial, the animate and inanimate. But this "leaving behind" is no simple casting off; it is no simple matter of transforming matter.

The *apokalypsis* is upon us, finishing up the work begun in the Platonic pharmacy with its teleportation of Being. As the planet becomes, unevenly to be sure, virtualized, time in one of its dimensions has become speed, the seedbed of profit. Both Marx and Melville, in their prophetic roles, already understood this machinic compression of time that was gaining momentum in the mid-nineteenth century. Bartleby might have made his advent "motionlessly" at the door of finance and the law, but his quasi biographer persistently calls out to him, invisible but within the reach of the voice-pager, in the "haste of business" (Melville 1990, 9). And even at the end of the narrative, included already in its opening, the letters, undelivered with their messages of hope, "*speed* to death" (34). Our own period, which is beginning to recast periodicity, is one of the "deployment of techno-science or tele-technology . . . whose movement and speed prohibit us more than ever from opposing presence to its representation, 'real time' to 'deferred time,' effectivity to its simulacrum, the living to the non-living, in short, the living to the living-dead of its ghosts" (Derrida 1994, 169).

The intent of the most "advanced" social formations of the transepochal is to increase transaction speeds, to reach for the appearance-vanishing of the instantaneous, to such an extent, that the turbines of the virtual will suspend death itself, and cause it to shimmer so ferociously that it will begin to disappear, to lose its uncanny, ghostly presence, and to submit itself to the protocols of the techno-capitalistic determination of objects. Then it can be rendered superfluous, cast aside so that the project of mastery can proceed unabated, without obstacles and without friction. Cyberspace, i ek observes, generates a fantasy of "a frictionless flow of images and messages—when I am immersed in it, I, as it were, return to a symbiotic relationship with an Other in which the deluge of semblances seems to abolish the dimension of the Real" and Bill Gates has proclaimed an era of "frictionless capitalism" (i ek 1997, 156). Everything should run as smoothly as the oiled music of ball bearings in motion, but there is always a hitch.[2]

Being-suspended—which is not presence, absence, or either's lack—is the essential attribute of our transepochal epoch. As Gianni Vattimo has noted: "The riskiness that belongs to the game can be seen in the fact that liberation for the exclusive bond to its historical context puts *Dasein* itself in a state of suspension; a suspension which touches him in his deepest constitution as subject . . ." (1993, 128). And this suspension strains so forcefully that it begins to oscillate, "The sphere of oscillation to which thought accedes by responding to the call of *Gestell* is a sphere in which metaphysics, and in particular the nature/history distinction founded on the schema of historiography and physics are suspended" (182). Nature becomes history; history becomes nature; and both are virtual echoes that we hear as if muffled, as if our ears and eyes are changing their structure and need hearing aids, telescopes, microscopes, oscilloscopes, and particle accelerators in order to see clearly and to hear above the din of the technical itself. "The essence of time lies in the ecstatic unitary oscillation . . ." (Heidegger 1984, 208), but the oscillation, generated by the friction of times (organic, cosmic, machinic, memorial, imaginal, etc.) is now moving so rapidly that we now exist strung out along all the old lines. "In its last endeavor, the thought of Heidegger," Briault insists, "is neither positive nor negative, nor even indifferent to affirmation or negation. Rather it is *suspending*" (cited in Rapaport 1989, 258). And Heidegger's thought is suspenseful because the showing forth of Being in its contemporary mode of technologics has suspended the entire human project.

In this time of suspension, time is being reconsidered. "From a post-Heideggerean perspective," Herman Rapaport argues,

> time is anything but a linear movement of the history of being or meaning; rather time is a manifold of relations in which the difference between moments is itself undecidably given in a trace structure in whose indeterminancy the various modalities of time (arche, moment, lapse, eschaton, duration, present, past, future, suspension) are given not simultaneously, but also not unsimultaneously. (65)

If neither simultaneously nor unsimultaneously, then the "now" and its others must be thought of differently than as the presence of the present. The "unitary" aspect of the ecstatic oscillation is being deconstructed and the line—such an elegantly simple, elegantly illusory concept of history—explodes and unravels, opening up into the multidimensionality of the "manifold relations" of the worked givenness of *language*. The essential question of *temps* moves from the definitional "what-is?" to the poetics of the "how shall we say?"

Freud, as we have seen, stands at one of the primal crossroads of (post)modernity, accomplishing the double tasks of unifying the time of the oedipal subject, history, and language, *and*, simultaneously, disarticulating the subject, history, and language from the constraints of a fantasy of univocal identity, thus releasing them toward the multiplicity of otherness (which he calls the unconscious, drives, and death). In a series of compositions analogous to those of the dream work, Freud accomplishes this paradoxical event by caulking the seams between antiquity and (post)modernity, between the primeval and the contemporary, between nature and culture. "Antiquity" and "(post)modernity," which he conjoins as the technocratic age of Oedipus, exist only as traceworks of signs, and it is only via the caulking itself that two such epochs come into being and are given the possibility of fitting or not fitting into a narrative. There is no antiquity, (post)modernity, or transepochality until we say these into being, time and again, through the symbolic networks of language (taken in its broadest sense as multiple signifying practices).

In the transepochal moment that has been actively prepared as the fulfillment and the countermeasure (the antidote?) to a philosophy that, doubling itself without respite, counts always on counting, time begins to temporalize itself in a different language, a language of difference. It is, to once again use Derrida's translation of Freud, a *flickering*, an oscillation that acts to suspend the usual denominations of past-present-future as a successive series analogous to a line. The switchings on the tracks that have kept the train of time for so long in a forward, progressive motion are snagged, stuck, and there is a *jolt* of planetary proportions occurring.

The temporal train, that which we thought was automobility itself, of course keeps moving and at an ever faster clip—these are, after all, ancient autobahns—but there is now, nonetheless, the suspicion that everything has somehow, and at the same time, become derailed, thrown off track. The *train à grande vitesse* is moving so fast that the image blurs and the sound of the human scream fades away into the distance. The trains have long ago jumped the tracks, and while it is true, as I have attempted to show from the Platonic teleportation onward, that "this 'derailment'—this lack of support, of a fixed instinctual standard, in the co-ordination between the natural rhythm of our body and its surroundings—characterizes man *as such*: man *as such* is 'derailed'" (i ek 1997, 135), the contemporary moment radicalizes the clattering roar, with unforeseen consequences.

Technologics moves at such a speed, far beyond the "natural" speed of the organic body, that the phonematic conversion from event into word and syntax cannot occur, except as an always belated response. And this very observation, the fact that all of these temporalities are occurring simultaneously, simply illustrates the multiple fields of time in which human life now occurs. It is, as Heidegger has phrased it, a "tearing together [*reißt zusammen*]."[3] As time tears along, it tears apart, an enigmatic movement that conjoins unexpected crossovers, the organic and its others, for example, into an uncanny jointure. Time, in the transepochal crossing, is a new form of dream time that is dissolving the fantasy of linearity even as the effects of that linearity continue to develop, unabated and turbocharged. It is, simultaneously, the *completion* of the primordial wish for the domination of the earth and death through the instantiation of the rational procedures that constitute technicity *and* the disruption of those procedures and their rational grounds. No one can predict how long—since "length" itself is one of the categories in question—either the completion or the disruption will require. Dream time normally shows itself as a spectrally fluid combination of image, words, and affect that calls for interpretations—a translation, in psychoanalytic terms, from the primary process into the more rational language of the secondary processes—that unravel the knots of the dream, thereby moving it along, but always without exhausting the potentiality of the dream.

There has always been the navel of the dream and of the body: the sign of dependency, a cut, and the blind spot of (un)readability. The dream occurs as a surrealist symbolic text, associative and magical, a rending and gathering of the unusual, without transitions or clear boundaries, and one can "begin" with any of the elements in order to initiate the work of reading. The temporal countertradition of the dream unfolding is not governed by the logic of identity explicated by Aristotle or the Platonic step-grid with its orderly hierarchy of knowledge. In the sense we are trying to fathom, in the transition between the ancient cultural dreamtime and contemporary technotime, the dream gets its wish of the transfiguration of the world into a world that can resist time's effects, but only by giving up its poetics, its functions of an ultimately resistant symbolization, of something-(not)there that *cannot* be read in the mode of explication. The navel (with its connection to the mother and natural birth) is where explanation falls silent, abandoning its resources, which are incommensurable to the task.

After asserting that "There is no meaning without some dark spot, without some forbidden/impenetrable domain into which we project fantasies which guarantee our horizon of meaning" (1997, 160), i ek argues that modern science, at least at the level of theoretical potentiality, abolishes that "dark spot," thus destroying the function of meaning for *Dasein*. "The outcome of the *suspension* of the dark spot of Beyond in the universe of modern science," he continues,

> is thus that "global reality" with no impenetrable dark spot is something accessible only on screen: the abolition of the phantasmic screen which served as the gateway into the Beyond turns the whole of reality into something which "exists only on screen," as a depthless surface. Or, to put it in ontological terms: the moment the function of the dark spot which keeps open the space for something for which there is no place in our reality is *suspended*, we lose our very "sense of reality." (163; emphasis mine)

For technotime, everything must become readable and writable code without remainder: the navel of the dream, that point which disappears into (and opens up) darkness, must be destroyed if possible, and, if not, then radically repressed. Everything, without exception, must be brought under the domination of digitization, the profits of the bottom line, and measurable utility.

This is a vast simplification of the processes by which such a transition has occurred; after all, this is the very history of the world. Every eventuality occurs in the most subtle gradations, through interactions that are finally untraceable in their complexity. All of this—as is any reading—is only one possible schema that interrogates the simplest contours of the *silhouette* of the textuality of temporality.

The transepochal moment of the suspension of animation approaches its completion as we see the shift from a *figure* (as face, diagram, shape, and trope) of reason to what can only be called, in all its embarrassing awkwardness, the *fact* of reason, its being embedded in every form of the lifeworld. Objectification is a sign of this reason, and both capitalistic work and the evolution of human beings have required just such an objectification of the world, making it into Heidegger's "standing-reserve" with all of the implications of this development. This long practice of objectification includes the objectification of the person as subject. The subject becomes an object of inquiry and of commodification and thus we see the emergence of the human sciences, of capitalism, and of things such as markets in organs and genetic codes. Accompanying this objectification, however, has been the concurrent project of the humanizing of the thing, of the animation of objects. This can be observed in the translation of nature into culture in the designed environment, in the long fascination with puppets and statues that come to life, with automated toys, and, later, with robots and contemporary questions about intelligent machines.

Many (post)Enlightenment commentators support the former tendency, the objectification of the world and the subject as a salubrious achievement of the sciences, while deriding the latter as "animism," "magical thinking," "anthropomorphizing," or as an egregious example of that worst of all human traits, the "pathetic fallacy." All of these interpretations depend upon a teleology that Freud, as only one representative, names as the progression from magic, through religion, to science (which corresponds with infancy, adolescence, and adulthood and their respective types of mentation: magic, a mixture, and rationality). From this perspective, organic activity is considered "living" and the rest of the natural world is considered "dead," animated only by the projection of human desire.

There are, however, other possibilities for reading the phenomenology of various "regions" of the world. "Now, what if Others were encapsulated in Things, in a way that Being towards Things were not ontologically severable, in Heidegger's terms, from Being toward Others?" asks Avital Ronell. She continues:

> What if the mode of *Dasein* of Others were to dwell in Things, and so forth? In the same light, then, what if the Thing were a *Dublette* [double/duplicate] of the Self, and not what is called Other? Or more radically still, what if the Self were in some fundamental way becoming a Xerox copy, a duplicate, of the Thing in its assumed essence? This perspective may duplicate a movement in Freud's reading of the uncanny [in "The Sandman"], and the confusion whirling about Olympia as regards her Thingness. Perhaps this might be borne in mind, as both Freud and Heidegger situate arguments on the Other's signification within a notion of *Unheimlichkeit*, the primordial being not-at-home, and of doublings. (1991, 24)

Just as the so-called natural world is humanized, so, too, the human is naturalized, thingified if you will, but without the negative connotations usually associated with "thing" (as dead, inert, nonconscious, commodified, etc.). If the "human" is changing its designation, so, too, will all the other terms associated with it.

The transepochal occurs when these two enormously powerful forces intersect, thus creating the conditions for the suspension of the (in)animate. The subject becomes object and the object becomes subject. Nature and culture, the *physis* of the self-blooming and that which is "artificially" constructed by human beings, are now becoming hybridized so that the time of nature and the time of culture are becoming inseparable for the mutants that we are. The time of technologics is the time of, and for, mutation.

Time is pluralized, a movement that dissolves enlightenment teleologies and hierarchies of magic, religion, and science and creates blended social forms that partake, as in a surrealist collage, of all older forms simultaneously. Analyzing contemporary culture, Davis concludes that "Magic is technology's unconscious, its own arational spell. Our modern technological world is not nature, but augmented nature, super-nature, and the more intensely we probe its mutant edge of mind and matter, the more our disenchanted productions will find themselves wrestling with the rhetoric of the supernatural" (1998, 38). As I have shown in my discussion of Freud's construction of the Oedipus, this strategy of interpretation also breaks down the traditional (for a certain history) genres of history such as "antiquity" and "postmodernity." The digitization of the world makes the entirety of the past, insofar as it has left traces, available in a blinding flash of the present, even as the usage of the past-present-future lineage is also spinning vertiginously. (Reading, with eyes closed and fingers gently running across the page, attempts to darken this flash, to slow it down to the speed of comprehension.)

The doubling of the Platonic-oedipal age of *mimēsis* gives way to the infinite multiplication of the age of electronic reproduction. Instead of two Cratyluses, instead of the self-object and its doubled other in the realm of ideas, or of the dyad of original copy, in the transepochal anything (including the self) that can be translatable into the electronic can be reproduced and distributed in enormous numbers, as long as there are receiving stations at hand. Time, rather than subsisting as part of the time/eternity duality, becomes pluralized, polychronic, and multivalent, and requires that we, as mutants still working to attend to the hyphenated *Da* of *Da-sein* and *fort-da*, learn to make linkages between disparate times without a false attempt to unify these times into the time of the subject, the object, the thesis, the universe, God, or knowledge.

Since time is multiplied, so, too, is the ghost. The revenant returns, now, from everywhere at once. The ghost no longer simply haunts the grave site, the castle, the house, the side of the road; rather, it haunts the world as world. "A specter is haunting Europe" has been globalized; Elsinore (with its multitude of questions of power, succession, love, melancholia, fathers and mothers) appears as every TV screen, every phone jack, every wireless computer. The world is shimmering, oscillating at unheard(-of) speeds and we are experiencing the being unhomed entailed by the absence/presence of the *Unheimlichkeit*.

This still movement flows across all the delineations of the old-fashioned ontological grid; the specter is thing, animal, object, commodity, automaton, the "animated-inanimated thing, the dead-living thing is [also] a father-mother" (Derrida 1994b, 152). We are all temps, transients, without a home and therefore haunting the streets, strung out along the arteries and networks, living by our witz. The uncanny haunts us in the texts of Freud, Heidegger, and Derrida, who are themselves driven by doubling. As Weber has pointed out, "The figure of the double is one of the most persistent leitmotifs in Derrida's writings. . . . The double is the ghostlike manifestation of *iterability*, which . . . 'splits' each element while at the same time 'constituting' it in and through the split" (1996, 144).

This "ghostlike manifestation" occurs in the closest intimacy with death, the self, and—as a form of iterability—*mimēsis*, a doubling that is bivalent. "Thus, on the one hand, repetition is that without which there would be no truth. . . . Here repetition gives itself out to be a repetition of life. . . . But on the other hand, repetition is the very movement of non-truth: the presence of what is gets lost, disperses itself, multiplies itself through mimemes, icons, phantasms, simulacra, etc. . . . Here, tautology is life going out of itself beyond return. Death rehearsal" (Derrida 1981, 169).

This bivalence, Derrida contends, can also be found in the writings of Heidegger, where there is "at least in principle, a kind of division between *hypomnēsis* and *anamnēsis*, with technique and writing on one side and poetic thinking on the other—in short, a bad and a good writing" (Derrida and Ferraris 2002, 8). Perhaps such a general bivalence is inevitable, produced by language and the writing subject taking a position, any position, within writing: by writing the first word. But it need not be reified into stark binary oppositions. In the play of the language of all the philosophical stylists, as well as in literature as literature, the plurality of bivalences can all be honored, developed, cast into the eddies of textualities.

In Freud's well-known essay on the uncanny focused on his reading of E.T.A. Hoffmann's "The Sand Man" (1817), there are automatons with limited vocabularies, the son's dread of the death of the father, narrative perplexities, and repetitions of the language of fire. Everything, when there is a passion for the speechless and feminized passion machine, is a burning. Neil Hertz, in his appropriately titled *The End of the Line*, points out that the essay deals with Freud's relationship to (not) literature and its fundamental concerns: the wish to be original, the fear of plagiarism, and the rivalry among writers. "The impulse to rewrite 'The Uncanny'," he continues, "may have been Freud's wish to test the value of his theory . . . but it might also have been his exclamatory response (*"Unheimlich!"*) to the theory's strangeness" (1985, 99). In fact, the text's "particular structural complexity—a temporal lag which produces, retroactively, a situation in which a text cannot be characterized as unequivocally 'real' or unequivocally 'fictitious'—is remarkably close to that of Freud's own notion of the workings of what he termed *Nachträglichkeit* in conferring meaning and pathogenic power on infantile experiences and fantasies." (107). This equivocalness between the fiction and the real, and its relation to the complex of questions strung through the guide wires of repetition, resonates at every intersection of the network of the transepochal.

The lines are indeed crossed in this conjunction between story and essay, the one doubling the other—Freud quotes only dialogue, thus playing the role of ventriloquist much like Socrates in the *Crito* and Plato throughout the dialogues—as the placement of the "literary" and the "analytic" does not remain at all clear and distinct. While the *theory* of psychoanalysis certainly attempts, like rationality itself, to *apply* itself to a concrete enigma and *explain* the inexplicable, it also carries forward, as a repetition, the bewilderment, perhaps the terror, of the experience of the uncanny as that which others.

These symptoms, however, are no longer carried primarily by neurotic individuals or neurotically inflected texts—although these still bear symptomatic burdens—but also appear as cultural formations in a globalized domain of simulacra. Being unhoused means more than leaving the lovely *Vorhof* of the mother's womb, for now *Dasein* as *Dasein* is being unhoused from the organic, especially the body itself, and from the earth. As Ronell puts it as she traces Heidegger's phone calls: "Technology is no tool and it no longer has anything to do with tools, but it provides an uncanny deracinating grid whose locality is the literalization of the *Unheimlich* . . ." (1991, 39). The *Gestell* is haunted, but, since this is "literalized" it tends toward invisibility. For the linkage between the *Unheimlich* and technologics to move from the invisibility of the ideological and everyday to the visibility necessary for any possibility of working-through, a poetics, a series of readings, must set itself into action. It, of course, always already has.

In *Being and Time*, Heidegger argues that uncanniness is constitutive of *Dasein*'s Being-in-the-world. The not-at-home is more primordial than the at-home (1962, 234), but for the most part *Dasein* covers up its uncanniness with the familiar, the everyday, the public voice of the "they." The mood of anxiety, however, brings us "face-to-face" with uncanniness, which, in turn, serves as the caller of the call of conscience in the "uncanny mode of *keeping silent*" (322). Uncanniness is "soundless" (343), but, nonetheless "The caller is *Dasein* in its uncanniness: primordial, thrown Being-in-the-world as the 'not-at-home'—the bare 'that-it-is' in the 'nothing' of the world" (321). The uncanny, playing a role structurally analogous to the Platonic *daimon*, links the "is" and the "not," being and nothingness.

The call illuminates the possibility of authenticity and "points *forward to*" (325) *Dasein*'s possibilities. Uncanniness, as it were, not only surrounds *Dasein* on all sides—as its foundationless foundation, as its own "having-been," and as its ownmost possibility of Being-toward-death—but also permeates *Dasein* to the "core" (if there were a core other than the relational structure of care), thus "individualizing" it. When *Dasein* is in its own truth, uncanniness is its condition of being. Everything is enigmatically lit; and, in a certain way, Being itself is the uncanny: unhomed and unhoming. Usually, however, *Dasein*

> entangles itself in itself so that not tarrying becomes *never-dwelling-anywhere*. This latter mode of the Present is the counter-phenomenon at the opposite extreme from the *moment of vision* [*Augenblick*]. In never dwelling anywhere, Being-there is everywhere and nowhere. The moment of vision, however, brings existence into the Situation—and discloses the authentic "there." (398)

And the "moment of vision" is a mode of temporality, for it is "Ecstatical temporality [that] clears the 'there' primordially. It is what regulates the possible unity of all *Dasein*'s existential structures" (402). The momentous blink of the *Blick* that gives *timing*, rather than the *logos* or the Kantian transcendental imagination, is what grants unity to *Dasein*'s Being-in-the-world.

What, though, does all this have to do with the technological and its affirmation and disruption by the transepochal crossing? It seems as though the uncanny serves as a direct line from *Dasein* to *Dasein*, from one to one. We are individualized by the uncanny as it calls us to our ownmost possibilities, which, on one hand, are the factical choices we make in our own Situation, and, on the other, the certainty of our being headed, always already, toward death. Such a direct line, however, assumes a certain linearity of time, which it is no longer possible to valorize as the essence of time.

Heidegger himself has radically problematized this line with his analysis of the temporalizing of the *ecstases* of time, and Derrida has, even more extensively, developed the traces of the lack of presence of the present. As he writes in *Writing and Difference*:

> The structure of delay (*Nachträglichkeit*) in effect forbids that one make of temporalization (temporization) a simple dialectical complication of the living present as an originary

and unceasing synthesis—a synthesis constantly directed
back on itself, gathered in on itself and gathering—of reten-
tional traces and protentional openings. The alterity of the
"unconscious" makes us concerned not with horizons of
modified—past or future—presents, but with a "past" that
has never been present, and which never will be, whose future to come will never be a *production* or a reproduction in
the form of presence. (1978, 21)

Examples of this extremely condensed critique of Hegel-Husserl-
Heidegger could of course be multiplied, but this one does nicely to
act as a "circuit breaker" of *Dasein*'s uncanny linkage with itself.
Derrida reads the Freudian "unconscious" as a metaphysical name
of this delay, and Ronell, explicating this same passage, clarifies that
"This is not a hidden, virtual or potential self-presence but an apparatus that sends out delegates, representatives, proxies, phony messages, and obscene calls taken but not essentially put through, often
missing their mark" (1991, 85).

If there were no delay, no relay systems that along their line of
servers act to mediate *Dasein*'s (self)-consciousness, then there could
be no phenomenon called "haunting" in which the other returns. Simultaneously from the past and from the future. That is, there could
be no experience of the uncanny itself (and therefore, finally, no in-
dividuation, no structure of care, no encounter with the nothing, no
Dasein). The delay, the stutter-step of time, serves as the necessary
condition for the appearance of the uncanny existence of existence.[4]
Edward Casey has pointed to the importance of this figure for Derrida's critique of presence, which rests on an interpretation of the
present as the privileged moment of time:

> For the decisive instance of his deconstruction of presence in
> his early work is precisely *the blink of the eye*. It is in this blink
> that any claim to the unmediated presence of forms as well as
> that of the "now" is exploded by a momentary cut that belies
> the primacy and privilege of such pure presence: "Nonpres-
> ence and nonevidence are admitted into the blink of the in-
> stant. There is a duration of the blink, and it closes the eye.
> This alterity is in fact the condition for presence, presentation,
> and thus for *Vorstellung* in general." (1999, 81)

The delay, the shutter-blink of the *Augenblick*, gives *Dasein* the
worlding of the world.

The desire of technologics, like that of the dual inheritors of rationalism called capitalism and science, is to obliterate the delay—everything should be instantaneous and always the newest of the new—and, thus, the uncanny and its pluralized proxies. The just-in-time inventory of absolute information should be always at hand, at the click of a button, at my beck and call, so that there can be a quick turnaround. Waiting should be outlawed. Exchange should occur at the speed of business, and time should be used up, utilized, numbered. 24/7. Desire should be satisfied before I am aware that I am desiring (since desire, too, has for so long involved a structure of delay, substitution, and deferment). It should be rationalized, quantified, properly channeled, so that its uncanny qualities (related to memory and anticipation, to mortality and the ethics of finitude, to the doublings of fantasy) might be reigned in, harnessed for the good of the bottom line.

In addition to the destruction of the wait, there are other indices of the relationship between the uncanny and technologics. For the most part, as we have seen, *Dasein* is that entity which, distracted from itself by objects, "never-dwells-anywhere." And if it dwells without a "where," it is also without a "when." Time passes; "I" float in its neutral medium. Contentedly ferried along the boulevards of the ideological matrix of the they, *Dasein* is anything *but* individualized by the (non)presence of the uncanny. (As both Gertrude Stein and William Gibson have said, "There is no there there.") There is a clearing, of sorts, but it is experienced only as a screen on which images sent by others are unreflectively absorbed and acted out. There is no wait, no imaginal twisting of the received impressions, no enigma about that which is sent and received. There is no mediated reply, no ability to respond with the prolonged effort necessary for a working-through. *Dasein* cannot be located, cannot be placed.

This is analogous to the nonplaceability of the technological itself, which is also everywhere and nowhere, without center or periphery, since it is a set of linkages always in motion. There are, of course, specific instantiations of technics as process and object, but the worldview that produces the examples is most often kept under the cover of the "obvious," "the way things are." This is Heidegger's *Gestell*, a word variously translated as "enframing," "installation," and, by Weber, as "emplacement," since

what is at stake is not the placing of something but the staking out of place as such. . . . As emplacement, the goings-on of modern technics thus display a markedly ambivalent character: they arrest, bring to a halt, by setting in place; but this placement itself gives way to other settings, to the incessant re-placing of orders through which new places are set up and upset. (Weber 71–72)

This "ambivalent character" of the *Gestell* suspends everything even as it moves with ever greater speed and in unprecedented currents. Technologics (dis)places. It slides us into slot machines even as the machine is being reconfigured. It imprints us with identity even as the imprint is being redesigned, phased out in place of newer forms of currencies.

Wondering whether the "ultimate objective of technics is a certitude and a security that is unwilling to leave any place to *incalculability* and chance" (53), Weber notes that "It is this ambivalence between the desire to occupy a place of one's very own and the desire to break out of a place in which one is caught, that is raised to *incalculable* proportions . . ." (6; italics mine). It is the incalculable that enables the aporias of the calculability of emplacement. If technologics is to destroy chance, it will have to become the absolute master of space-time and obliterate death, traditionally the prerogative of God. As the ancient dream of immortality and the will-to-power of the logic of technicity are more and more tightly wound around each other and become a cable, the incalculable comes under the sway of calculability and Zapparoni will be able to manufacture as many bees, automatons, or Cratyluses as he desires (although this would altogether recalibrate desire).

There is, of course, a countertradition to the tradition of counting all things one after the other. For both movements, which are not disconnected from one another, different things count in different ways. Accountability is understood differently, even if everybody is a prestidigitator, playing with finger magic. The veil is merely a prop for the magician. In the well-known passage from *Being and Time*, for example, Heidegger writes that:

If the question of Being is to have its own history made transparent, then this hardened tradition must be loosened up, and the concealments which it has brought about must be dissolved. We understand this task as one in which by *taking the question of Being as our clue*, we are to destroy the traditional content of ancient ontology until we arrive at those primordial experiences in which we achieved our first

ways of determining the nature of being—the ways which
have guided us ever since. (1962, 44)

Without again rehearsing the history of the deconstructive work of
Nietzsche-Freud-Heidegger-Derrida (which I read as an affirmation, a
"yes, yes"), let me make a few comments on the above passage.

I take it that the twentienth century, with its many modernisms
and their "posts," has loosened, if not dissolved, the congealments of
history, and that certain "primordial experiences" have been, if not arrived at (which is structurally impossible), at least more closely ap-
proached. By, say, a step. From another direction, both "primordial"
and "experience" have been called radically into question by this same
deconstructive practice. Even more fundamentally, the "we," the "our,"
and the "us"—along with their companions, the "they" and the "I" —
have begun to oscillate. If, in yet another simplification, we take "ancient ontology" to be tantamount to metaphysics, and this, in turn, as
tantamount to technologics, then as the tradition has been completed
it has also released its other(s) that are the shadows and ghosts of the
technologized world. All hangs suspended.

Dasein is not itself—and it is never a thing, present-at-hand—
without the structure of time, which Heidegger will analytically
translate as "temporality."

> Coming back to itself futurally, resoluteness brings itself into
> the Situation by making present. The character of "having
> been" arises from the future, and in such a way that the fu-
> ture which "has been" (or better, which "is in the process of
> having been") releases from itself the Present. This phenom-
> enon has the unity of a future which makes present in the
> process of having been; we designate it as *"temporality."*
> Only in so far as *Dasein* has the definite character of tempo-
> rality, is the authentic potentiality-for-Being-a-whole of an-
> ticipatory resoluteness, as we have described it, made
> possible for *Dasein* itself. Temporality reveals itself as the
> meaning of authentic care. (1962, 374)

It is futurity, as care and as being-toward-death, that constitutes
the "essence" of *Dasein*. In an extreme twist of our everyday understanding of time, the past arises from the future, without
thinking of the future as the next point on the time line, but as
what Heidegger calls the *"primary phenomenon of primordial and
authentic temporality"* (378).[5]

Weber asks what might be required not to take the past for granted, to rethink the structure of time. "First and foremost," he writes:

> it might require the present to rediscover its own heterogeneity by becoming more sensitive to the way in which the past exists not merely as a derived form of the present, as a present which "is" no more, as a *past present*. The past might be approached not simply as a weak or deficient mode of the present but rather as a dimension and function of that iterability which belongs as much to the future as to what we commonly think of as the present. Iterability belongs more to the future and to the past than to the present because it never comes full circle, never comes to rest in a simple, straightforward identity. What iterability entails is not a simple return of the same but a process of alteration and transformation that involves difference no less than identity. (1996, 148)

Iterability, as we have seen, is neither good *mimēsis* nor bad *mimēsis,* but the enabling condition of all (re)production. Iterability is a movement of constitutive divisiveness, an originary proliferation, and therefore in addition to "enabling" *mimēsis* it also simultaneously disenables it. With its fantasy of cloning—the perfect reproduction of the same—*mimēsis* is broken up from the so-called beginning. And since *mimēsis* is part of a flowered wreath that includes identity and time in its arrangement, these, too, are pluralized. Time times time; identity is always more than one. There are voices that surround us, inhabit us; there are ghosts that wander across all demarcations, that make the past and future present in the multiply figured present that undoes the unity of the present.

All lines are entangled. Technologics, which laid down the law of the line, is now unthinkable without the crossings and crossovers it is generating. Our present, then, is an extremely bizarre event in which boundaries and frames (such as those around "event") are coming undone. As the rational drives itself—it is, after all, the law of automobility—more and more deeply into its others (the irrational, chance, the materiality of the body, the empirical as a whole), it seems as if the uncanny, at last driven away by the calm and objective utility of the rational machine, would simply *evaporate.* Technics would show the uncanny, its own ghostly apparition, to be just that: an apparition, with no "basis" in "reality."

Instead, the uncanny—the sense of strangeness—is, for those listening, being massively amplified as we exist more and more as those whose being is dislocation. Earth, since the first technology and the inception of *Homo* (as *erectus*, *habilus*, *floreiensis*, *sapiens*: take your pick) has been a crossing place, a zone of transition that serves as the (dis)location for the othering of animal bodies, including the human. As the animal and the automaton settle ever more deeply into the tissue circuits of the other, and as the doubleness of *mimēsis* is replaced by the multiplicity of the transepochal means of (re)production, so, too, the figure of the human and its putative other, the monster, is being transformed.

If Prometheus, Faust, and Frankenstein can be said to be the figures that most fully articulate the emergence of modern technologics in the nineteenth century, then those rather lonely and titanic figures have now morphed. From their still recognizably human form, they have now been disarticulated and displaced into the nodes of every social system; they have been *incorporated*, in both the psychoanalytic and economic senses, into all the circulating networks of information. And if their electrified, computerized bodies, having been de-corporalized, haunt the interior of a body without organs, then a certain mourning must occur that, *perhaps*, can come to take the form of anticipation that does *not* reject death. I'm not counting on it.

This is a repetition of the dismemberment of the Titans and the Gods, but with a difference: it is an undoing powered by the nucleus and the electron, it has other forms of transmission than narrative and ritual—that is, it is often, and more often, transmitted from machine to machine—and it is ever-expanding, rather than cyclical. Osiris-Orpheus-Dionysos: nature torn, scattered, and recovered in a sacred *sparagmos*. But now, in the midst of the construction of a global electronic exoskeleton, there is no longer the tragedy of the dithyrambs or the lyricism of the sonnets of Orpheus, but only a low steady humming that only rarely articulates itself in nontechnical language. Prometheus, Faust, and Frankenstein have broken out of the frames of their fictional texts and now move anonymously, without a proper name, along the grid lines of the *Gestell*: through laboratories and the offices of politicians, along the streets of noir with their neon storefronts, among the professors at their workstations, through the phone lines, and into the interstices of the chromosomes.

It is no accident that the monsters have taken to the streets, to the screen, to the family room, and into "outer" space. Speaking as a psychoanalyst, André Green has written of the fragmentation of time.

Thorough analysis shows that the analysand addresses him/herself to an object which is not only not utilizable through its absences, but is, in fact, connected to a *phantom* of the object that is swarming with vampiric desires. This object that, to a large extent, bears the responsibility for the infant's incapacity to surmount traumatic situations, has nevertheless in the long run become a narcissistic object without any relations with the object of past reality; a sort of monstrous creation, with its maliciousness, its indifference, its incomprehension and from which one is able to detach oneself only by treating oneself in the same manner: cruelly, with an insensitivity to suffering, and closed off from all understanding. (2000, 153)

In our culture of the simulacrum, with its connections to borderline disorders and the sadomasochism of narcissism, the monstrous, as the vampiric and otherwise, plays a pivotal role as a barely conscious expression of terror as well as a projective defense against the same. "This is why," Ronell explains, "the technological flower of the Frankensteinian project becomes so crucial, as the objectivization of the replacement object, a high-wired monument to its impossible switch . . ." (1991, 222). No such "objectivization," of course, can be entirely satisfactory, although desire works mightily to find its fulfillment through the technological drive. And just because of its failures, the monstrous is acted out through a serialized repetition that always entails murder. And, as Ronell further observes, the "switch"—of monstrosity, of compulsion, of serialization, and of technologics itself—does not exist. Technology cannot be turned on and off.

We, then, *are* the monstrous—selving the other and othering the self—that is traversing every boundary. If, terrified of our unrecognized subjectivities, we destroy the monster, the game's up. The monster can be neither deconstituted nor domesticated, but it can be, at least partially, named. It can be addressed. We can learn, or begin to learn, a new language for our (post)human condition. We can listen, with our old ears and with our new, refined prosthetic hearing devices. We can respond to what the monster demonstrates, to what it shows forth into being. This is a frightening process, and we must submit to being haunted, to telling ghost stories, and to the fact that the time of our time is undergoing an earthquake, being radically disrupted and wrested away from the linearity of a progressive, history and from narratives of self-completion in this world or the next.

Time is being disengaged from the line that has, for so long, governed it, a metamorphosis that is difficult in the extreme to describe. And it is less a description, with the implications of a subject outlining an object already in place, than a bringing-to-pass through language. This difficulty arises for all the reasons that Derrida has so assiduously presented about the "contamination" of the language of metaphysics and about the impossibility of suddenly waking up outside the closure of that tradition. There is no inside and outside; there is only the (re)working of language, again and again.

Working-through is this reworking of textuality in its expansive sense that opens up, ever so slowly, a reworking of temporality. Language, passing through its own history, provides the shadowed light that might, if only barely, shine on that which has been covered over. And, since technologics is unthinkable outside the history of *mimēsis*, and these, in turn, are convolved with the relation of subject to object and all the metaphysical notions of the unity of the self, then to work on any one of these elements—a labor that can only occur in and through language, although perhaps best along its Möbian borders— is to reconfigure them all.

It is to reconfigure the configuration itself.

As we attempt to understand our position within the turbulent dynamic of the crossings of *technē* and *physis*, we must attempt to understand a "here" and a "there" that are not stable, that provide no fixed location for thought. We must think on our feet, on the move, and there is no elsewhere toward which we might flee to escape the predicament of the vanishing sea and the loss of directionality. It is this caughtness, this being irremediably snagged, that creates the conditions for the suspension of being in which we forge ourselves. We exist along the jagged edges, not yet smoothed to a black matte finish, between technologics and a thinking that attempts to think against the grain of technologics, attempts to think the other of metaphysics. Some of the words that like shooting stars appear and disappear in this zone are the eternal return, Zarathustra, the unconscious, dreamwork, *Gelassenheit*, the fourfold, thanking-thinking, *différance*, the promise, the event, the inhuman, justice, and friendship.

This thinking-writing does not negate the technological. Indeed, as Derrida has rightly insisted: "No one is allowed on these premises if he is afraid of machines and if he still believes that literature, and perhaps even thought, ought to *exorcise the machine*, the two having nothing to do with each other" (1981, 292). There is no thought, no experience, no culture, no literature—and there never has been—without machines at work. This thought does, however, use every prosthetic and supplementary device at its disposal in its attempt to listen from within to the rumbling whir of the technicity of existence and to that temporality which is a kind of dynamic *stalling*.

These thinkers, using every resource of language that they can invoke, *gesture toward*. As Emmanuel Levinas has suggested, time "signifies this *always* of non-coincidence, but also the *always* of *relationship*, an aspiration and an awaiting, a thread finer than an ideal line that dia-chrony does not cut. Dia-chrony preserves this thread in the paradox of a relationship that is different from all the other relationships of our logic. . . . Here there is a relationship without terms, an awaiting without an awaited, an insatiable aspiration" (1987, 32). The noncoincidence of time, an expression of temporalizing differentiality, draws us into an unsurmisable, incalulable relationship with the world that, quite mysteriously, maintains a fine line of continuity. Time gathers, with tears, and between the inspiration and the expiration is the "insatiable aspiration": the held breath, suspended, of an attention that works to wait upon.

Conclusion:
Heeding the
Phantomenological

Higher than love of the neighbor is love of the farthest
and the future; higher yet than the love of human beings I
esteem the love of things and ghosts. This ghost that runs
after you, my brothers and sisters, is more beautiful than
you; why do you not give him your flesh and your bones?

—Zarathustra

Perhaps, though, in the end nothing has changed. After all, we
all wake up in the morning, go about our business as we live
out our numbered days, and then slip away, at one time or another, into sleep at first and then, at last, into the night of nothingness. We continue to hope for a long life, for love in its many forms,
and for a graciously courageous death. We are anxious about being
absorbed into the Borg, the hive-mind of technology, and we want
technology to produce faster computers and more precise medi-
cines. The wish is still alive, and we all willingly participate in it.
Postmodernity, one name for the opening to the posthuman, is not
all that bad for those of us with luck enough to have money for
housing and food. We'll just continue to limp along, as human beings always have. With the advent of technologics, nothing has
changed—nothing fundamental, at any rate. As I have tried to
demonstrate through readings of Platonic *mimēsis*, the "prefer not"
of Bartleby, the immortality machine of Marx, the logic of the drone,
the installation of surveillance from Oedipus to Lacan, and the

twisting of temporality in the transepochal, to be human has always been to be posthuman, the machinic has always been embedded within the organic. From a certain perspective, then, one that we might still call that of the line and the point, we are where we have always been. Walking in place.

But the logic of the line and the point, as the basic trope of time, history, causality, knowledge, and identity, is precisely what I hope to have called into question. At each "stage" of the development of technologics—a thinking pragmatics that wants to peel the name and the mind away from the body, that wants all things to be disposable as calculable utility, and whose deepest dream is to abolish death—there has also always been the presence of the disrupting and multiplying effect of ghosts, those absent ones who nonetheless make demands on us and of us. We have never been where we have thought ourselves to have been; we have always been elsewhere and other than now. The human has always been structured as a structure of displacements and deferrals. Mutation, therefore, can occur. As Baudrillard reminds us, "[I]n relations between things there is always a hiatus, a distortion, a rift that precludes any reduction of the same to the same" (2000, 71).The phantomenological must supplement the phenomenological, with its historical links to the appearing of the clear daylight of consciousness.

Things have always been the same, and the same has always been troubled by ghosts, riven by wights that wait, that speak, that demand an attempt at justice. These are two general conclusions to be drawn from this discussion of technologics. But in the transepochal something *is* changing, something different *is* emerging. We can all feel it; we are trying to articulate it.

The posthuman is the crisscrossed moment during which the most primordial of dreams is being realized empirically, when what has been operative in the ideal realm of thought, as forms of repetition, comes to be instantiated in an externalized form as a cyborg. The earth is coming under the dominion of the electronic, everything is being mediatized, and the elements of the (in)organic are changing places. What is to come is not predictable. The posthuman is not something given, programmable, a conclusion of the linear sequence of cause-and-effect. How could it be? It is, at least for a moment, an open possibility to be shaped by an untold complexity of forces, including but not limited to intentionality.

The posthuman is not—not necessarily—the antihuman as a dystopian vision of absorption into the machine, as the cataclysmic end of humanity in which freedom, ethics, and vocation vanish forever, although this is one of the many possible trajectories the species could take. Katherine Hayles and Jean Baudrillard are among those who have laid out their double vision for the future of the posthuman. Hayles outlines her logic of the future by telling us that:

> If my nightmare is a culture inhabited by posthumans who regard their bodies as fashion accessories rather than the ground of being, my dream is a version of the posthuman that embraces the possibilities of information technologies without being seduced by fantasies of unlimited power and disembodied immortality, that recognizes and celebrates finitude as a condition of human being, and that understands [that] human life is embedded in a material wold of great complexity, one on which we depend for our continued survival. (1999, 5)

I tend to agree with her view of the situation, except that I am suspicious of the division of the visionary into the dream/nightmare binary, as this will again sort itself through the Platonic-monotheistic register of values. We will have, simultaneously, both versions of the posthuman at work in the experiment we are undertaking.

After laying out her narrative of how information has lost its body, the creation of the cyborg after World War II, and the transition from the human to its next stage, Hayles concludes that for the posthuman, "emergence replaces teleology; reflexive epistemology replaces objectivism; distributed cognition replaces autonomous will; embodiment replaces the body seen as a support system for the mind; and a dynamic partnership between humans and intelligent machines replaces the liberal humanist subject's manifest destiny to dominate and control nature" (288). This "partnership," which will never be absolutely symmetrical, finds other political-economic paths than domination; the teleological line from origin to end transforms itself into a multifoliate bouquet; consciousness is only one of many important subprograms of the human, and thus a new modesty might be engendered; and embodiment is understood as the sine qua non of life, posthuman or not. Each of these arrays of practices will have to be worked and reworked as we construct the culture of the posthuman.

Baudrillard, as well, sees two fundamental possibilities. The first is that the domination of technology, with its rage to overcome death, will enact what he calls the "perfect crime," creating a completely smooth world that will obliterate the human as well as any evidence of any crime having been committed. It will be the event that erases all eventuality, including the "illusory world's definitive opacity and mystery" (2000, 74), or what he, along with many others, calls simply the "secret." That would, indeed, be the end of the human, beyond which there would be no singularity of thought even though (re)production and a simulacra of the human would continue on unabated.

There is, however, another chance, and for Baudrillard the alternative would be that

> at the extreme horizon of technology, something else happens, another game, with other rules. The point is that the constellation of the secret still resists, remains alive. Either we think of technology as the exterminator of Being, the exterminator of the secret, of seduction and appearances, or we imagine technology, by way of an ironic reversibility, might be an immense detour toward the radical illusion of the world . . . an absolutely unpredictable movement that would finally bring us to the other side of metaphysics. (82)

The "other side of metaphysics" would be a metaphysics radically and unpredictably transfigured, and one that would, opening, hold the secret close, perhaps even with love. And to prepare for this other thinking, which may well never arrive—perhaps, as Derrida has suggested, it is in its very nature not to be able to arrive as any empirical event—all we can do is to recycle, make compost piles, cut down on our road rage, continue to destroy the planet, read books, use our cellphones judiciously, and throw out a few lines of poetry. It's not much, but it's a chance.

In *A Taste for the Secret*, Derrida reminisces about one of his longstanding fantasies. "When I was very young—and until quite recently," he writes, "I used to project a film in my mind of someone who, by night, plants bombs on the railway: blowing up the enemy structure, planting the delayed-action device and then watching the explosion or at least hearing it from a distance" (Derrida and Ferraris 2002, 51). Deconstruction, apparently, is the bomb. As a result of being bombed, not only will the enemy's life become more hazardous, but the "friend, too, will have to live and think differently, know where he's going, tread more lightly" (52). All of us, friend or foe, are film works, cinematic projectors. Structures are blown up by night, and at night, as we drift into sleep; technologics produces a delayed-action device that, although it saturates our every move, is usually heard or seen from a distance. And in the very midst of this series of incessant, often silent, explosions, we must learn to live differently.

What, asks Avery Gordon, is a ghost? "It is a case of haunting," she replies to herself, to her listening other, "a story about what happens when we admit the ghost—that special instance of the merging of the visible and the invisible, the dead and the living, the past and the present—into the making of worldly relations and into the making of our accounts of the world" (1997, 24). Unless we are able and willing to address the ghosts by which we are addressed—to converse with the phantasms which haunt our world both psychically and socially—then we will continue to be driven by them into acts of violence, into the madness of forgetting (which is not the same as the necessity for forgetting), and, if worst comes to worst, to the smoothing out of all the snags of existence that mark the human as human, whether *post-* or not.

My hope is that we can learn to dance with the terrible beauty of the arabesque formed by the flaming sword of the angel without, out of terror and the wish to expunge death, attempting to douse the flame and step back into the garden. The garden is gone; we must learn to make-do in the wilderness, in the shadowed darkness of the cave that opens out onto the weather of the earth. This is our dwelling: a house of dreams, of simulacra, of phantasms that will not let us rest. Technology will of course continue to develop, and this is the fundamental adventure of the posthuman. And, yet, if we pay heed to technologics not just as the measurable *ratio* of rationality, with its criteria of the clear and distinct, but also as a discourse of hauntology, then perhaps the networks of the technical will open out onto new possibilities and that "other metaphysics." We cannot know; we can only work.

And this work, the work of forging the symbolic within, and perhaps just to the side of, the space of technologics will not be frictionless. It will tear. It will be a halting work, step by step, and a work that often loses whatever ground it has gained. There will be laborers who unearth the secrets of the past, dating it with atomic half-lives and ordering its stratifications, in order to return to the earth. There will be star watchers and particle tracers. There will be those who stand alone, waiting, and those that stand in the midst of the cosmopolis, spending freely the capital of political imagination. But for all who prepare by working within the symbolic, there will be the structure of a delay. The Parousia will not come, for it is always yet-to-be, and that is the ground for writing, for again learning to read.

The delay is the blink, stutter, stutter-step, vacillation, oscillation, gap, detour, lapsus, spacing, iterability, punctuation, gulp, echo, roundabout, suspension. Events that we can somehow manage, even if they break us, and the nonevent of death, which "I" will not experience and that cannot be managed, even though we fear it and mourn for it, futurally, almost constantly. This is an opening onto the posthuman that refuses the homogeneity of spaces, times, and lines, that follows the adventure of technology with the greatest curiosity, while always listening for the undertones that indicate the presence of the restless shades. We are, after all, the last men and women, called into our own being by a Nietzschean discourse on overcoming, on suspension as an act of bridging other shores while keeping them apart. As the figures of the end, dazed and traumatized, what can we do but blink?

"What does that mean?" Heidegger asks. "Blink is related to Middle English *blenchen*, which means deceive, and to *blenken*, *blinken*, which means gleam or glitter. To blink—that means to play up and set up a glittering deception which is then agreed upon as true and valid—with the mutual tacit understanding not to question the setup" (1968, 74). This blinking, in an immense condensation of Heidegger's thought on technology, as if it could all fit in a nutshell, is a "consequence of a type of representation already dominant" (83). So we are the ones who know that all that glitters is not gold and agree not to question this comfortable arrangement with deception. But we are also other than this, the ones for whom blinking takes on other cadences, for blinking can also be understood as a form of the delay, the interruption of vision that is necessary for any seeing. It enacts the lack of complete presence in the present; it adds a "spectacular blinking-effect to the warning light . . . [and] is a rhythm essential to the mark" (Derrida 1988, 188–89), whose functioning Derrida repeatedly analyzes with such lucidly obsessive passion. It enables the rhythm of spacing and timing, and therefore of a subjectivity that concomitantly entails a range of understandings of objectivity. No longer, however, will the so-called subject and the so-called object be *opposed* to one another; rather, they will come to be experienced as different modes of the appearance of time-space, both in incessant motion.

Along with many others, Guattari has begun to suggest what tasks might be entailed as the machine-assemblages that we are cobble together a preparation for posthuman culture. Noting that the "polyphony of modes of subjectification actually corresponds to a multiplicity of ways of 'keeping time'" (1995, 15) and that this polyphony is the "mutant rhythmic impetus of a temporalisation able to hold together the heterogeneous components of a new existential edifice" (20), he then proceeds to argue for the importance of a poetic function in the posthuman world; its task would be "to recompose artificially rarefied, resingularised Universes of subjectivation . . . of catalyzing existential operators capable of acquiring consistence and persistence" (19).

How are we, though, to "keep time" with the posthuman? Tap our prosthetic feet and snap our digitized fingers? As we mutate, time, too, will continue to take on new contours, only one of which will be linearity, and all of which, if *Dasein* is to continue as *Dasein*, will depend on the preservation of the delay. Instant gratification, with its destruction of historicity and futurity, would be disastrous. Rather, we must keep time by keeping faith with the tasks of thought—as they summon us—at least while we are left with a "disjointed or dis-adjusted time without which there would be neither history, nor event, nor promise of justice" (Derrida 1994b, 170). This is a time of technologics that recognizes itself as a pragmatics and a discourse haunted by a host of specters that require a hearing, rather than the turbocharged domination of the earth that imagines rationality without remainder, without unincorporable otherness as a constitutive aspect of itself.

This task will involve an array of resistances to the powers that set everything up to be remainderable so that we are left without remains. Exploitation of a variety of sorts, as Negri reminds us, is "real and intolerable, and we can [fight] this only in constituting a new reality, a new hybrid being, different each and every time, constructed and therefore snatched away from humanity's arch-ghosts with each instance" (1999, 14). A malleable, always emerging hybridity, yes, but there will always be ghosts with which to contend, with which to wrestle in order to formulate a new name. There will never be a *time* of the total exorcism of the "arch-ghosts," a time beyond the "vital illusions" of ideology or mythology as collective belief. I concur with Derrida when he remarks that he "would not free [him]self so easily of phantoms, as some people all too often think they do" and with his claim that we are "structured by the phantasmic and in particular that we have a phantasmic relation to the other, and the phantasmicity of this relation cannot be reduced, this pre-originary intervention of the other in me" (Derrida and Ferraris 2002, 89). Phantasmicity, in other words, is the necessary condition for any experience of the human and its *post-*.

The ontology of the clear and the distinct, the fantasy of a purely rational methodology that will forever unveil the face and the body of the world, has given way to hauntology. And phenomenology, the sunstruck science of the appearance of appearances, must also, in the transepochality of the posthuman, be complemented by phantomenology, the (non)science of dusk and the night. Each formation has its temporality, and every temporality has its language. El, Yahweh, Christ, the *eidos* and the *chora*, the human-machine of capitalism, the prefer-not, the eternal return of Dionysus and Apollo, the hyperspeed of the drone, *Nachträglichkeit*, *Ereignis* and the fourfold, iterability and the *a-venir*. Thinking the temporality of the posthuman requires the ongoing invention of a new cultural encompassing, of hybrid subjectivities, a magical science, a politics of serious play, and of a language that turns the old language inside out in order to open another edge. A new mechanico-animism, a new mythotechnics. A poetics of the cyborg.

This, of course, has already gotten underway. Technologics as phantomenology asks that we invite the ghosts into our solar-powered living rooms, and that we, for our part and as lovers of the unprogrammed future, leave home for those places of trauma and desecration that are in need of our slow attentiveness. The phantoms, to be sure, cannot be domesticated, but they can, if honored, perhaps bring their obscure blessings, their asymmetrical justice, to the posthumans that we are becoming.

The Countdown

We're done, for now. Go home, if you can.
To repeat the program, push enter.
Reset.
Reloop the text.
It will run again, saming difference,
othering the same.
Guaranteed,
(for the
time (of) (not) being).

All writing is ghost-writing, ghosts writing to ghosts about ghosts, traces of ash and smoke on the wind.

The living and the dead cross at the threshold, gazing upon each other as they silently pass. As a narrator named for a place reflects in a novel-memoir of the same name:

It does not seem to me that we understand the laws governing the return of the past, but I feel more and more as if time did not exist at all, only various spaces interlocking according to the rules of a higher form of stereometry, between which the living and the dead can move back and forth as they like, and the longer I think about it the more it seems to me that we who are still alive are unreal in the eyes of the dead, that only occasionally, in certain lights and atmospheric conditions, do we appear in their field of vision. (Sebald 2001, 185)

But such a spacing of time occurs only in the field of writing, itself a form of spatial temporality, of self-differentiation. Writing on behalf of another, the other who calls for writing, by writers whose names are obscured and marked through by the words they write. Credited, yes, in some way, but with a name eclipsed by other needs.

The dead are silent. They speak, calling.

They've got our number. We are commanded to stay on the line, waiting—

The past is yet to come; the future is so far behind us that we cannot, even by turning our heads, see it emerging.

Push redial, then call-forwarding.

Listen for the tone—

Will you pick up? If so, what will you say? How will you answer this call?

The ghostly voices, speaking as if in a *chora*, a chorale, one voice after the other, blend into what might be called a tone, but not one of harmony, not quite one: a tone indicating a next step in a *danse macabre*, the sound of static that moves in place and in turmoil, *suspended*, that acts something like a handshake—

Notes

INTRODUCTION

1. For a brief survey of the increase in the usage of "ghost" in everyday life, see McQuain 1999.

CHAPTER 1: –CALL FORWARDING

1. See Hillis, 1999a for the details of the relationship between gaming and the military.

2. For philosophical exegeses of voice and call, see Heidegger 1962 (vol. 2 pars. 54–60), Derrida's analysis of the Heideggerian *Stimme/Stimmung* in Derrida 1983; Agamben 1991; and Ronell 1991. Heidegger asks "How is the conscience to function as that which summons us to our ownmost potentiality-for-Being, when proximally and for the most part it merely *warns* and *reproves*?" (1962, 324). For Freud, the still small voice functions as both warning *and* call, as we shall see, from the *tyrannos*.

3. For a summary of the legal case revolving around the use and ownership of John Moore's spleen, see www.richmond.edu/~wolf/moore.htm, provided by Michael Allan Wolf, and Boyle 1996. As Boyle says, correctly, "behind the visible information revolution in technology and economy, a significant but unexamined process of rhetorical and *interpretive* construction is going on. This process of construction produces justifications, ideologies, and property regimes rather than mainframes, software, or gene splices. Yet it will shape our world as thoroughly as any technical change" (ix). It is the structure of this "interpretive construction" that I am attempting to examine.

Chapter 2: –On-(the)-Line

1. For a general discussion of the surveillance society in which we live, and more specifically for an argument against closed-circuit cameras as constant observers of everyday life in the city, see Rosen 2000.

2. The translation is from Aristotle 1983.

3. For Hegel's presentation on space-time, see Hegel 1970 §253–61.

4. For discussions of the relationship between Jünger and Heidegger, see Losurdo 2001, esp. 192–206, as well as Zimmerman 1992 and Bullock 1992.

5. Here and in chapter 3, all quotations from the *Republic* are from Plato 1992. All quotations from the *Apology*, *Cratylus*, an *Phaedo* are from Plato 1985.

6. I am quite taken with train stories, including those by Sebald (in whose work trains are never *not* present), Hrabal, and Schivelbusch. I have recently discovered that Louis Althusser, in the mid-'70s, began to focus on the metaphor of trains, which was to "become one of the main preoccupations of his last years" (1997, 9). He himself remarked that "There is no point of departure and no destination. One can only ever climb aboard moving trains; they come from nowhere and are not going any place. Materialist thesis: it is only on this condition that we can progress" (10).

7. Green (2000) discusses an entire range of the temporalities of psychoanalysis that he calls "heterochronicity." I will let *Nachträglichkeit* resonate with them all.

8. Hayles discusses the differences between "artificial intelligence" (AI) and "artificial life" (AL), arguing that the goal of AI was to build, "inside a machine, an intelligence comparable to that of a human. The human was the measure; the machine was the attempt at instantiation in a different medium . . . by contrast, the goal of AL is to evolve intelligence within the machine through pathways found by the 'creatures' themselves. . . . Whereas AI dreamed of creating consciousness inside a machine, AL sees human consciousness, understood as an epiphenomenon, perching on top of the machinelike functions that distributed systems carry out. In the AL paradigm, the machine becomes the model for understanding the human. Thus the human is transfigured into the posthuman" (1999, 239). I will, in

general, not make these distinctions and will use "posthuman"—as Hayles usually does—in a larger context than simply the product of the AL model.

CHAPTER 3: –THE PLATONIC TELEPORT

1. In his *Confessions*, Augustine says of the angels' reading practices: "Let the peoples above the heavens, your angels, praise you. They have no need to look up to this firmament and to read so as to know your word. They ever 'see your face' (Mt 18:10) and there, without syllables requiring time to pronounce, they read what your eternal will intends. They read, they choose, they love. They ever read, and what they read never passes away. By choosing and loving they read the immutability of your design. Their codex is never closed, nor is their book ever folded shut" (1992, 283). There is a slippage of the word "read" as it moves toward the immediate intuition of the angels, but, in any case, those of us who are nonangelic require the *time* of reading and an unfolding of the book.

2. Weber 1986, on Derrida's place, follows the logic of moving from the (non)readability (of the term *chez*), to accountability (of the *chez*), to a consideration of time and that which orders space-time.

3. From Pygmalion to Frankenstein and beyond, the duplicated comes to desire individuality. For recent examples, see, for instance, films such as *Blade Runner*, *Bicentennial Man*, *AI*, and the host of stories in which the robot takes on individual consciousness (along with mortality, morality, desire, and all those other human qualities). And, since we are all preprogrammed to begin with, this applies to each of us as well.

4. This is not just an ancient philosophical fantasy. As Hazel Muir has written about the work of Dmitry Sokoloff, a physicist who, along with a number of other researchers, has been looking for a "ghost" of the Milky Way: "[I]f you travel in a straight line through a finite Universe, you will see the same pattern of stars and galaxies repeated time and again, as if you were moving through an endlessly repeating set of identical regions of space. . . . But what if the alien galaxy is just an image of the Milky Way, and what if the aliens are us? This is exactly what could happen if the Universe was made of identical, repeating copies of itself. . . . With an infinitely repeating pattern of ghost images there would be many such rings . . ." (Muir 1998).

5. For history of "noise" in cybernetics, see Davis 1998; Hayles 1999; and Taylor 2003.

Chapter 4: –The Elixir of Life

1. There is a rich genealogy of the philosophical-psychoanalytic interpretation of money. For a review of this literature, see Weschler 1999a.

2. Derrida says in a note that "According to Freud, we adopt our judgement to the conditions of *fictive* reality, such as they are established by the poet and treat 'souls, spirits, and specters,' like grounded, normal, legitimate existences. A remark that is all the more surprising in that all the examples of *Unheimlichkeit* in [his] essay are borrowed from literature!" (1994, 196).

3. Another doctor, this one named Frankenstein, notes with a similar tone: "And now my wanderings began which are to cease but with life. I have traversed a vast portion of the earth and have endured all the hardships which travelers in deserts and barbarous countries are wont to meet. How I have lived I hardly know; many times have I stretched my failing limbs upon the sandy plain and prayed for death. But revenge kept me alive; I dared not die and leave my adversary in being" (Shelley 1991, 185). Also, the "soaring" Winzy mentioned is constitutive of a certain form of the wish for transcendence that includes Platonism, many forms of monotheism, Gnosticism, and those technodreamers who work toward the separation of the meat of the flesh from the informational patterns of identity. It is visible everywhere, including the final shot of *The Matrix*. For a discussion of these connections to the high-tech drives, see Davis 1998.

Chapter 5: –The Immortality Machine of Capitalism

1. In Fellini's *Juliet of the Spirits*, the protagonist beautifully works through her superego and its guilt. At the film's end, with the cast taking its leave in that Felliniesque signature moment, Juliet walks out of a gate and onto the sunlit grounds of her estate. The body, in Brown's sense, is resurrected. And yet we are still within the frame of art and its production via illusion. There is liberation, but it is only into a wider, more spacious, *frame*. This is perhaps the goal of all symbolic action in its sense of "working through."

2. In one example of how such idealized capitalist time is broken apart, Michel de Certeau examines what in "France is called *la perruque*, 'the wig.' *La perruque* is the worker's own work disguised as work for his employer . . . [and] may be as simple a matter as a secretary's writing a love letter on 'company time' or as complex as a cabinetmaker's 'borrowing' a lathe to make a piece of furniture for his living room. Accused of stealing or turn-

ing material to his own ends and using the machines for his own profit, the worker who indulges in *la perruque* actually diverts time (not goods, since he uses only *scraps* [my emphasis]) from the factory for work that is free, creative, and precisely not directed toward profit. In the very place where the machine he must serve reigns supreme, he cunningly takes pleasure in finding a way to create gratuitous products whose sole purpose is to signify his own capabilities through his *work* and to confirm his solidarity with other workers or his family through *spending* his time in this way. . . . Though elsewhere it is exploited by a dominant power or simply denied by an ideological discourse, here order is *tricked* by an art" (1984, 25–26).

3. Marx quotes from Goethe, *Faust*, pt 1, act 5, the scene in Auerbach's cellar. The lines occur as the drinkers are singing the song of the poisoned rat when Mephistopheles and Faust enter.

4. Robert Reich describes these workers as "symbolic analysts" who can "solve, identify, and broker new problems" (1991, 103–4).

5. Marx's syntax is puzzling, disjointed, and repetitious, but the translator provides a straightforward explanation: The first sentence in the quotation is "inserted by Marx, above the line, *in English*; thus the apparent virtual repetition. (The sentence following also appears in English in the original)" (1973, 292 n. 4). English above the line; German on the line; then English, again on the line, to conclude. English is a voice-over that repeats the German; or is it the other way around? Isn't this the structure of the "virtual repetition"?

Chapter 6: –Bartleby the Incalculable

1. Commodification, indeed all of capitalism, is a form of social magic, or to use another register, of spiritualism. Derrida analyzes all of this with his usual panache in Derrida 1994b. See also Cutler 2002, esp. 48–54 and Weinstein's "Technologies of Vision," where she demonstrates the links between spiritualism and technology. As she writes: "While individuals and machines had not yet completely amalgamated into one concept, in mid-nineteenth-centure America, they nonetheless were being thought of as less and less distinct from one another. The trope worked both ways: As science became more and more sophisticated, human beings began to embody a biomechanics whereby persons took on values and abilities of inanimate machines; likewise both physiologists and engineers began to speak of machines as possessing a vital, human nature" (Weinstock 2004, 130).

She makes the intriguing suggestion that following the invention of the telegraph, the daguerreotype, and the railroad, the "inherent chaos of disembodied and panoramic communications and transportation that these media offer" was in part responsible for the fact that the United States began to standardize time. Jeffrey Sconce takes such considerations into the world of modern telecommunications in *Haunted Media*, where he responds to the question of why, "after 150 years of electronic communication, do we still so often ascribe mystical powers to what are ultimately very material technologies" (6), 2000.

2. Let me mark the topical news of the day. I am writing this as Enron, WorldComm, Qwest, Merck, Xerox, and so forth are coming under investigation for their "accounting" practices, which "shake the confidence" of the potential investors in Wall Street. No doubt, the Bartleby factor has been in place all along and is still in effect (and affect).

3. Derrida often speaks of the affirmative nature of deconstruction. This is particularly visible, for example, in "Ulysses Gramophone: Hear Say Yes in Joyce," where he remarks that "The minimal primary *yes*, the telephonic 'hello' or the tap through the prison wall, marks, before meaning or signifying: 'I-here,' listen, answer, there is some mark, there is some other. Negatives may ensue, but even if they completely take over, this *yes* can no longer be erased" (1992c, 298). Bartleby is a sign of this *yes*, within the very bereftness of his negation of signification.

4. An entire analysis of *Bartleby* could be done within the confines of *The Psychoanalysis of Fire*, beginning with, perhaps, "We should also note the loose association of the confused ideas of heat, food, and generation; those who wish male children 'will endeavor to nourish themselves with all the good, hot, and igneous foods'" (Bachelard 1964, 49). Bachelard explores, among other things, the sexual connections of electricity and fire.

5. The crowd always has a place in the work of the singular ones: Socrates, Kierkegaard, Nietzsche. Or of Melville himself: "[T]ake mankind in mass, and for the most part, they seem a mob of unnecessary duplicates, both contemporary and hereditary. But most humble though he was, and far from furnishing an example of the high, humane abstraction; the Pequod's carpenter was no duplicate . . ." (1954, 357). The carpenter is "unaccountable" and a "soliloquizer"; but no more of him right now, for he has a prosthesis to construct.

6. Continuing, for a moment, to follow in the wake of the whale, Ishmael speaks of the "tornadoed Atlantic of my being" (Melville 1954, 303), although he, unlike the scrivener, survives the wreck. There is a "Nippers," and according to Solomon, "'the man that wandereth out of the way of un-

derstanding shall remain' (i.e. even while living) 'in the congregation of the dead'" (ibid., 328). Who is dead? Who understands?

7. Storms appear often in the German tradition of weather watchers. Since we're in the neighborhood: "We see this lightning only when we station ourselves in the storm of Being" (Heidegger 1975, 78).

8. There is a "Cask of Amontillado" very close by. In Poe's story, published in 1846, there is a crypt within a crypt and, as Montresor says, "I reapproached the wall; I replied to the yells of him who clamored. I re-echoed, I aided, I surpassed them in volume and in strength. I did this, and the clamorer grew still" (Poe n.d., 266). One of the first reviewers of Melville's story said almost as much: "a Poeish tale, with an infusion of more natural sentiment" (*Literary World*, December 3, 1853, cited in Fiene 1966, 151).

9. A translator's note: "'Cathexis' is the generally accepted rendering of '*Besetzung*'. James Strachey coined the word in 1922 from the Greek, 'to occupy.' He records in the *Standard Edition* that Freud was unhappy with this choice because of his dislike of technical terms (S.E., III, 63, n. 2). The German verb '*besetzen*' is indeed part of everyday usage; it has a variety of senses, the chief one being *to occupy* (e.g. in a military context, to occupy a town, a territory). An alternative English translation, used occasionally, is 'investment', 'to invest'" (LaPlanche and Pontalis 1973, 65).

CHAPTER 7: —THE DRONE OF TECHNOCAPITALISM

1. This excerpt from the *Tischrede* is from a transcription, by Richard Brem, of a videotape of a television program broadcast live on the regional Südwestfunk (SWF) TV channel. It appeared on the Ernst Jünger listserv, ernst-juenger-1@maillist.ox.ac.uk (accessed Saturday, 21 September 1996 at 16:35:09). The translation is mine. I cannot number the communications systems involved that make it possible for me to hear this word and pass it along.

2. Jünger had a long-standing interest in the occult. For example, in a letter of August 27, 1922, he writes to his brother that to kill time during his yearly bout with the flu, he had some books sent over from the city archives that included some old "demonologies." He then goes on to compare these books to "old but still passable shafts of a mine" (cited in Mohler 1955, 68). Such texts take on, for Jünger, qualities of underground metallurgical work. Alchemy leads, through science and capitalism, to the cyborg. Or, as Dr. Angelo says in the film *Lawnmower Man*, "Once the goal of magicians and alchemists, we

can now use virtual reality to actually bring the human beings up to the next stage of evolution" (cited in Heim 1993, 143).

3. Beyond the manufacture and marketing of body parts, see the performance work of Orlan and Stelarc, as well as the self-experimentations of Kevin Warwick in the Department of Cybernetics at the University of Reading, England, who has had chips implanted in order to interact with the environment, both human and otherwise. All of these both critique and further the prosthetic body-computer relationship.

4. There is a vast literature on bees as symbols of divinity and spiritual work.

5. Of this brutally speeding car, Walter Benjamin has said that advertisement "abolishes the space where contemplation moved and all but hits us between the eyes with things as a car, growing to gigantic proportions, careens at us out of a film screen" (1978, 85). We are very close indeed—tailgating, as it were—to John Hawke's *Travesty*, to J.G. Ballard's *Crash*, and to Thomas Pynchon's *Gravity's Rainbow*, and even to the old simonized Chevvy of Arthur Miller's *Death of a Salesman*. Or see the charming photographs of Jacques Derrida, already on the road in his car, as a child (in Derrida and Bennington 1993). Car talk could go on indefinitely.

6. This scenario in which the fabricated is taken for the naturally given, one definition of ideology, is analyzed in Lacan 1977 in a discussion of the contest between Zeuxis and Parrhasios, ancient Greek painters.

7. For Jünger, J. P. Stern argues, the modern world is "die Werkstättenlandschaft," a landscape of workshops and scaffoldings (Stern 1953). (Cf. also Frank Lloyd Wright's, "our landscape of crude scaffoldings," cited in ibid., 41, 41 nn.)

Chapter 8: The Psychotelemetry of Surveillance

1. *Die Fuge* is translated "jointure," about which Heidegger writes: "This 'between' is the jointure in accordance with which whatever lingers is joined, from its emergence here to its departure away from here. The presencing of whatever lingers obtrudes into the 'here' of its coming as into the 'away' of its going. In both directions presencing is conjointly disposed toward absence. Presencing comes about in such a jointure" (1975, 41); or, "granted that the jointure thanks to which revealing and concealing are mutually joined must remain the invisible of all invisibles, since it bestows shining on whatever appears" (115). Invisible, perhaps, but nonetheless named; thus, thinkable.

2. Freud wrote on May 21, 94: "[I] am pretty well alone here in tackling the neuroses. They regard me rather as a monomaniac, while I have the distinct feeling that I have touched on one of the great secrets of nature" (1954, 83). And, on August 16, 1895, from Bellevue just before sending the "Project": "This psychology is really an incubus—skittles and mushroom-hunting are certainly much healthier pastimes. All I was trying to do was to explain defence, but I found myself explaining something from the very heart of nature" (1954, 123). There is much more to say about Freud and mushrooms. Weber begins his delightful discussion in the "Meaning of the Thallus": "If the dream-wish erects itself, phallic-like, out of the mycelium, the latter serves to remind us of what the Lacanian reading would like to forget: that the dream-navel cannot be reduced to a question of the phallus, of the *béance*, split or absent center of a subject, for one simple reason: the thallus" (1982, 81).

3. Yates argues that Grillparzer "suggests that he himself principally felt the kinship of his work with the popular *Geisterstücke* of his boyhood; but what the critics saw in the finished piece was a fate-tragedy" (1972, 50). Steiner agrees, noting that Grillparzer as a "dramatist of transition . . . sought to combine the Greek and the Shakespearean legacy into a form of tragic drama appropriate to the modern theater" (1980, 232). Both *Kindlers Literatur Lexikon* and Müller suggest a similar positioning of *Die Ahnfrau*. Benjamin, too, has little regard for the genre as a whole: "In respect of the forms mentioned, this reserve, indeed scorn, seems to be most justified where the fate-drama is concerned" (1977, 128).

CHAPTER 9: –TEMPS

1. It is beyond the scope of "Temps" to give a systematic interpretation of Heidegger's understanding of time as it is furthered and modified by Derrida. Heidegger 1962, Heidegger 1972, and Derrida 1982b are the texts always in the near background. For more thorough critiques, see Dastur 1998, Rapaport 1989, and Wood 1989.

2. In Derrida 1994a, there appears an editorial note: "Parts one to three of Jacques Derrida's answer is an editorial reconstruction of his argument due to a technical hitch in the recording" (21). Perhaps that is human history and all its representations: a technical hitch. So we have to hitch up our pants and get on with the reconstruction.

3. The term is from Heidegger 1971," but I came upon it in Weber 1996, 26.

4. Along another trajectory of conceptualizing the delay, Homi Bhabha has written that "the postcolonial passage through modernity produces a form of

retroaction: the past as projective. It is not a cyclical form of repetition that circulates around a lack. The time *lag* of postcolonial modernity moves *forward,* erasing that compliant past tethered to the myth of progress, ordered in the binarisms of its cultural logic: past/present, inside/outside. This forward is neither teleological, nor is it an endless slippage. It is the function of the lag to slow down the linear, progressive time of modernity to reveal its gesture, its *tempi*—'the pauses and stresses of the whole performance'" (1995, 58). The delay, for Bhabha, gives the possibility of the "in-between" that allows for cultural emergence.

5. Derrida questions the formulation: "And why qualify temporality as authentic—or *proper (eigentlich)*—and as inauthentic—or improper—when every ethical preoccupation has been suspended?" (1982b, 63). Dastur argues that these are not "ethical" categories at all, but formal possibilities of being-in-the-world that Heidegger draws from Husserl. (See Dastur 1998, 23 and 73 for her clarification of her point.) I cannot, however, see how *some sort* of hierarchical evaluation would not accompany this translation of *Eigentlichkeit-Uneigentlichkeit*. Once again, what most fascinates me is the prefix.

Works Cited

Agamben, Giorgio. 1991. *Language and Death: The Place of Negativity*, Trans. Michael Hardt and Karen Pinkus. Minneapolis: University of Minnesota Press.

———. 1993. *The Coming Community*. Trans. Michael Hardt. Minneapolis: University of Minnesota Press.

"*Anfrau, Die.*" 1986. In *Kindlers Literatur Lexicon*. München: Deutscher Taschenbuch Verlag.

Althusser, Louis. 1991. *The Spectre of Hegel: Early Writings*. Trans. G. M. Goshgarian. Ed. François Matheron. London: Verso.

Aristotle. 1983. *Physics III and IV*. Trans. Edward Hussey. Oxford: Clarendon Press.

Augustine. 1992. *Confessions*. Trans. Henry Chadwick. New York: Oxford UP.

Bachelard, Gaston. 1964. *The Psychoanalysis of Fire*. Trans. Alan C. M. Ross. Boston: Beacon Press.

Ballard, J. G. 1973. *Crash*. New York: Farrar, Straus and Giroux.

Bat-Ami Bar On. 1991. "Why Terrorism Is Morally Problematic." *Feminist Ethics*, ed. Claudia Card. Lawrence: University of Kansas Press.

Baudrillard, Jean. 1994. *Simulacra and Simulation*. Trans. Sheila Faria Glaser. Ann Arbor: University of Michigan Press.

———. 2000. *The Vital Illusion*. Ed. Julia Witwer. New York: Columbia UP.

Benjamin, Walter. 1978. "One Way Street." In *Reflections*, trans. Edmund Jephcott, ed. Peter Demetz. New York: Schocken Books.

———. 1977. *The Origin of German Tragic Drama*. Trans. John Osborne. London: NLB.

Bhaba, Homi. 1995. "Freedom's Basis in the Indeterminate." In *The Identity in Question*, ed. John Rajchman. New York: Routledge.

Borges, Jorge Luis. 1964. *Labyrinths*. Ed. Donald A. Yates and James E. Irby. New York: New Directions.

Bouwmeester, Dik, Jiai-Wen Pan, Klaus Mattle, Manfred Eibl, Harald Weinfurber, and Anton Zeilingen. 1997. "Experimental Quantum Teleportation." *Nature*, 11 December, 575–79.

Boyle, James. 1996. *Shamans, Software, and Spleens: Law and the Construction of the Information Society*. Cambridge, MA: Harvard UP.

Brotchie, Alastair, comp. 1995. *A Book of Surrealist Games*. Ed. Mel Gooding. Boston: Shambala.

Brem, Richard, transcriber. "The *Tischrede* of Ernst Jünger's 100th Birthday Party." Listserv: ernst-juenger-1@maillist.ox.ac.uk, on Sat, 21 Sep 96 16:35:09.

Brown, Norman. O. 1959. *Life against Death: The Psychoanalytical Meaning of History*. New York: Vintage.

Bullock, Marcus. 1992. *The Violent Eye: Ernst Jünger's Visions and Revisions on the European Right*. Detroit, MI: Wayne State UP.

Casey, Edward. 1999. "The Time of the Glance: Toward Becoming Otherwise." In *Becomings: Explorations in Time, Memory, and Futures*, ed. Elizabeth Grosz. Ithaca, NY: Cornell UP.

Certeau, Michel de. 1984. *The Practice of Everyday Life*. Trans. Steven Randall. Berkeley: University of California Press.

Cutler, Edward S. 2002. *Recovering the New: Transatlantic Roots of Modernism*. Hanover, NH: University of New Hampshire.

Dastur, Françoise. 1998. *Heidegger and the Question of Time*. Trans. François Raffoul and David Pettigrew. Atlantic Highlands, NJ: Humanities Press.

———. 2000. *Telling Time: Sketch of a Phenomenological Chronology*. Trans. Edward Bullard. London: Athlone Press.

Davis, Erik. 1998. *Techgnosis: Myth, Magic, and Mysticism in the Age of Information*. New York: Three Rivers Press.

Deleuze, Gilles. 1995. *Negotiations, 1972–1990*. Trans. Martin Joughin. New York: Columbia UP.

Derrida, Jacques. 1973. *Speech and Phenomena*. Trans. David B. Allison. Evanston, IL: Northwestern UP.

———. 1978. *Writing and Difference*. Trans. Alan Bass. Chicago: University of Chicago Press.

———. 1981. *Dissemination*. Trans. Barbara Johnson. Chicago: University of Chicago Press.

———. 1982a. *Margins of Philosophy*. Trans. Alan Bass. Chicago: University of Chicago Press.

———. 1982b. "*Ousia* and *Grammē:* Note on a Note from *Being and Time*." In *Margins of Philosophy*, trans. Alan Bass. Chicago: University of Chicago Press.

———. 1985. *The Ear of the Other: Otobiography, Transference, Translation.* Ed. Christie McDonald. Trans. Peggy Kamuf. Lincoln: University of Nebraska Press.

———. 1987. *The Post Card: From Socrates to Freud and Beyond.* Trans. Alan Bass. Chicago: University of Chicago Press.

———. 1988. *Limited Inc.* Trans. Samuel Weber. Evanston, IL. Northwestern UP.

———. 1992a. *Given Time: I. Counterfeit Money.* Chicago: University of Chicago Press.

———. 1992b. *The Other Heading: Reflections on Today's Europe.* Trans. Pascale-Anne Brault and Michael B. Naas. Bloomington: Indiana UP.

———. 1992c. "Ulysses Gramophone: Hear Say Yes in Joyce." In *Acts of Literature*, ed. Derek Attridge. New York: Routledge.

———. 1993. "Heidegger's Ear: Philopolemology (*Geschlecht* IV)." Trans. John P. Leavey. In *Reading Heidegger: Commemorations*, ed. John Sallis. Bloomington: Indiana UP.

———. 1994a. "Nietzsche and the Machine." *Nietzsche Studies* 7:7–66.

———. 1994b. *Specters of Marx: The State of the Debt, the Work of Mourning, and the New International.* Trans. Peggy Kamuf. New York: Routledge.

———. 1996. *Archive Fever: A Freudian Impression.* Trans. Eric Prevowitz. Chicago: University of Chicago Press.

Derrida, Jacques, and Geoff Bennington. 1993. *Jacques Derrida.* Chicago: University of Chicago Press.

Derrida, Jacques, and Maurizio Ferraris. 2002. *A Taste for the Secret.* Trans. Giacomo Donis. Eds. Giacomo Donis and David Webb. Cambridge, UK: Polity.

Fiene, Donald F. "Chronological Development, in Summary, of Criticism of 'Bartleby.'" 1966. In *Bartleby the Scrivener*, ed. Howard P. Vincent. Kent, OH: Kent State UP.

Freud, Sigmund. 1953. *On Aphasia.* London.

———. 1954. *The Origins of Psychoanalysis: Letters to Wilhelm Fliess.* Trans. Eric Mosbacher and James Strachey. New York: Basic Books.

———. 1969. *The Interpretation of Dreams.* Trans. James Strachey. New York: Avon.

———. 1974a. "Dissolution of the Oedipal Complex." In vol. 19 of *The Standard Edition of the Works of Sigmund Freud.* London: Hogarth Press and the Institute of Psychoanalysis.

———. 1974b. *Outline of Psychoanalysis.* In vol. 23 of *The Standard Edition of the Works of Sigmund Freud.* London: Hogarth Press and the Institute of Psychoanalysis.

———. 1989. *Civilization and Its Discontents.* New York: W. W. Norton.

———. 1991. *Die Traumdeutung.* Frankfurt am Main: Psychologie Fischer.

Gasché, Rodolphe. 1988. *The Tain of the Mirror.* Cambridge, MA: Harvard UP.

Gibson, William. 1984. *Neuromancer.* New York: Ace Books.

Goethe, J. W. 1964. *Faust.* Trans. Bayard Taylor. New York: Washington Square Press.

Gordon, Avery F. 1997. *Ghostly Matters: Haunting and the Sociological Imagination.* Minneapolis: University of Minnesota Press.

Goux, Jean-Joseph. 1993. *Oedipus, Philosopher.* Trans. Catherine Porter. Stanford, CA: Stanford UP.

Green, André. 2000. *Le temps éclaté.* Paris: Les Éditions de Minuit.

Guattari, Félix. 1995. *Chaosmosis: An Ethico-Aesthetic paradigm.* Trans. Paul Bains and Julian Pefanis. Bloomington: Indiana UP.

Hardt, Michael, and Antonio Negri. 2000. *Empire.* Cambridge, MA: Harvard UP.

Hawkes, John. 1976. *Travesty.* New York: New Directions.

Hawthorne, Nathaniel. 1959. "Dr. Heidegger's Experiment." In *The Complete Short Stories.* New York: Hanover House.

Hayles, N. Katherine. 1999. *How We Became Post-Human: Virtual Bodies in Cybernetics, Literature, and Informatics.* Chicago: University of Chicago Press.

Hegel, G. W. F. 1970. *Philosophy of Nature.* Ed. A. V. Miller. Oxford: Clarendon Press.

Heidegger, Martin. 1958. *The Question of Being.* Trans. Jean T. Wilde and William Kluback. New Haven, CT: College and University Press.

———. 1962. *Being and Time.* Trans. John Macquarrie and Edward Robinson. New York: Harper & Row.

———. 1968. *What Is Called Thinking?* Trans. J. Glenn Gray. New York: Harper & Row.

———. 1971. "The Origin of the Work of Art." In *Poetry, Language, Thought,* trans. Albert Hofstadter. New York: Harper & Row.

———. 1972. *Time and Being.* Trans. Joan Stambaugh. New York: Harper & Row.

———. 1975. *Early Greek Thinking.* Trans. David Farrell Krell and Frank A. Capuzzi. New York: Harper & Row.

———. 1977a. "The End of Philosophy and the Task of Thinking." In *Basic Writings,* ed. David Farrell Krell. New York: Harper & Row.

———. 1977b. *"The Question Concerning Technology" and Other Essays.* Trans. William Lovitt. New York: Harper.

———. 1984. *The Metaphysical Foundations of Logic.* Trans. Michael Heim. Bloomington: Indiana UP.

———. 1988. *Antwort: Martin Heidegger im Gespräch.* Ed Günther Neske and Emil Kettering. Pfullingen: Verlag Günther Neske.

———. 1996. *Hölderlin's Hymn "The Ister."* Bloomington: Indiana UP.

Heim, Michael. 1993. *The Metaphysics of Virtual Reality*. New York: Oxford UP.

Hertz, Neil. 1985. *The End of the Line: Essays on Psychoanalysis and the Sublime*. New York: Columbia UP.

Hillis, Ken. 1999a. "A Critical History of Virtual Reality." In Hillis 1999b.

———. 1999b. *Digital Sensations: Space, Identity, and Embodiment in Virtual Reality*. Minneapolis: University of Minnesota Press.

Hrabal, Bohumil. 1998. *Closely Watched Trains*. Trans. Edith Pargeters. Evanston, IL: Northwestern UP.

Janicaud, Dominique. 1994. *Powers of the Rational: Science, Technology, and the Future of Thought*. Trans. Peg Birmingham and Elizabeth Birmingham. Bloomington: Indiana UP.

———. 1997. *Rationalities, Historicities*. Trans. Nina Belmonte. Atlantic Highlands, NJ: Humanities Press.

Jünger, Ernst. 1957. *Gläserne Bienen*. Stuttgart: Ernst Klett Verlag.

———. 1960. *The Glass Bees*. Trans. Louise Bogan and Elizabeth Mayer. New York: Farrar, Straus and Giroux.

———. 1993. "On Danger." *New German Critique* 59 (Spring/Summer): 27–32.

Kant, Immanuel. 1965. *The Critique of Pure Reason*. Trans. Norman Kemp Smith. New York: St. Martin's Press.

Lacan, Jacques. 1977. *The Four Fundamental Concepts of Psychoanalysis*. Ed. Jacques-Alain Miller. Trans. Alan Sheridan. New York: W. W. Norton & Co.

LePlanche, J., and Pontalis, J. B. 1993. *The Language of Psychoanalysis*. Trans. Donald Nicholson Smith. New York: W. W. Norton.

Latour, Bruno. 1993. *We Have Never Been Modern*. Trans. Catherine Porter. Cambridge, MA: Harvard UP.

Levinas, Emmanuel. 1987. *Time and the Other*. Trans. Richard Cohen. Pittsburgh, PA: Duquesne UP.

Losurdo, Domenico. 2001. *Heidegger and the Ideology of War: Community, Death and the West*. Trans. Marella Morris and Jon Morris. Amherst, NY: Humanity Books.

Luckhurst, Roger. 1996. "(Touching on) Tele-Technology." In *Applying: To Derrida*. ed. John Brannigan, Ruth Robbins, and Julian Wolfrays. New York: St. Martin's Press.

Marx, Karl. 1939–1941. *Grundrisse der Kritik der Politischen Ökonomie*. Moscow: Marx-Engels-Lenin Institut.

———. 1973. *Grundrisse: Foundations of the Critique of Political Economy*. Trans. Martin Nicolaus. New York: Random House.

———. 1978a. *Capital, Volume One*. (Excerpts.) *The Marx-Engels Reader,*. ed. Robert Tucker. New York: W. W. Norton & Company.

———. 1978b. *Eighteenth Brumaire of Louis Bonaparte*. In *The Marx-Engels Reader*, ed. Robert Tucker. New York: W.W. Norton & Company.

McQuain, Jeffrey. 1999. "Ghost: The Haunting of Our Language." *New York Times Magazine*, 5 September, 28.

Melville, Herman. 1990. *Bartleby and Benito Cereno*. New York: Dover Publications.

———. 1954. *Moby Dick*. Boston: Houghton Mifflin.

Mohler, Armin, ed. 1955. *Die Schleife: Dokumente zum Weg von Ernst Jünger*. Zürich: Verlag der Arche.

Moravec, Hans. 1988. *Mind Children*. Cambridge, MA: Harvard UP.

Muir, Hazel. 1998. "Ghosts in the Sky." *New Scientist*. Archive: 19 September 1998. www.newscientist.com.

Müller. Joachim. 1964. *Franz Grillparzer*. Stuttgart: J. B. Metzlersche Verlagsbuchhandlung.

Negri, Antonio. 1999. "The Specter's Smile." In *Ghostly Demarcations: A Symposium on Jacques Derrida's "Specters of Marx*," trans. Patricia Dailey and Costantino Costantini, ed. Michael Sprinkler. London: Verso.

Newman, Lea Bertani Vozar. 1979. *Nathaniel Hawthorne: A Reader's Guide to the Short Stories*. Boston: G. K. Hall.

Nietzsche, Friedrich. 1966. *Thus Spoke Zarathustra: A Book for All and None*. Trans. Walter Kaufmann. New York: Viking Press.

———. 1971. *Joyful Wisdom*. Trans. Thomas Common. New York: Frederick Ungar.

Novalis. 1997. *Philosophical Writings*. Trans. Margaret Mahony Stoljar. Albany: State University of New York Press.

Peters, F. E. 1967. *Greek Philosophical Terms: A Historical Lexicon*. New York: New York University Press.

Pickering, Andrew. 1995. *The Mangle of Practice: Time, Agency, & Science*. Chicago: University of Chicago Press.

Plato. 1985. *The Collected Dialogues*. Ed. Edith Hamilton and Huntington Cairns. Princeton, NJ: Princeton UP.

———. 1992. *The Republic*. Trans. G. M. A. Grube, revised by C. D. C. Reeve. Indianapolis, NJ: Hackett Publishing Company.

Poe, Edgar Allan. N.d. "The Cask of Amontillado." In *Tales*. New York: Grosset and Dunlap.

Rapaport, Herman. 1989. *Heidegger and Derrida: Reflections on Time and Language*. Lincoln: University of Nebraska Press.

Reich, Robert. 1991. *The Work of Nations*. New York: Knopf.

Ronell, Avital. 1991. *The Telephone Book: Technology, Schizophrenia, Electric Speech*. Lincoln: University of Nebraska Press.

Rosen, Jeffrey. 2000. *Unwanted Gaze: The Destruction of Privacy in America*. New York: Random House.

Sconce, Jeffrey. 2000. *Haunted Media: Electronic Presence from Telegraphy to Television*. Durham, NC: Duke UP.

Sebald, W. G. 2001. *Austerlitz*. Trans. Anthea Bell. New York: Random House.

Shelley, Mary. 1981. *Frankenstein*. New York: Bantam.

———. 1996. "The Mortal Immortal." In *Literature by Women: The Traditions in English*, ed. Sandra M. Gilbert and Susan Gubar. New York: W. W. Norton & Company.

Smock, Ann. 1998. "Quiet." *Qui Parle* 2, no. 2 (Fall): 68–100.

Steiner, George. 1980. *The Death of Tragedy*. New York: Oxford UP.

Stern, J. P. 1953. *Ernst Jünger*. New Haven, CT: Yale UP.

Taylor, Mark C. 2003. *The Moment of Complexity: Emerging Network Culture*. Chicago: University of Chicago Press.

Terramorsi, Bernard. 1991. "Bartleby or the Wall." *Revue Littéraire Mensuelle* 69 (April): 87–99.

Ulmer, Gregory. 1985. *Applied Grammatology: Post(e)-Pedagogy from Jacques Derrida to Joseph Beuys*. Baltimore: Johns Hopkins UP.

Vattimo, Gianni. 1993. *The Adventure of Difference: Philosophy after Nietzsche and Heidegger*. Cambridge, UK: The Polity Press.

Virilio, Paul. 1991. *The Aesthetics of Disappearance*. Trans. by Philip Beitchman. New York: Semiotext(e).

Weber, Samuel. 1982. *The Legend of Freud*. Minneapolis: University of Minnesota Press.

———. 1986. "Reading and Writing—*chez* Derrida." In *Institution and Interpretation*. Minneapolis: University of Minnesota Press.

———. 1991. *Return to Freud: Jacques Lacan's Dislocation of Psychoanalysis*. Trans. Michael Levine. Cambridge: Cambridge UP.

———. 1996. *Mass Mediauras: Form, Technics, Media*. Stanford, CA: Stanford UP.

Weiner, Susan. 1992. *Law in Art: Melville's Major Fiction and Nineteenth-Century American Law*. New York: Peter Lang.

Weinstein, Sheri. 2004. "Technologies of Vision: Spiritualism and Science in Nineteenth-Century America." In *Spectral America: Phantoms and the National Imagination*, ed. Jeffrey Weinstock. Madison: University of Wisconsin Press.

Weinstock, Jeffrey. 2003. "Doing Justice to Bartleby." *ATQ: 19th Century American Literature and Culture* 17, no. 1 (March): 23–42.

Weschler, Lawrence. 1999a. "Bibliographic Essay." In Wechsler 1999b.

———. 1999b. *Boggs: A Comedy of Values*. Chicago: University of Chicago Press.

Wood, David. 1989. *The Deconstruction of Time*. Atlantic Highlands, NJ: Humanities Press International.

Yates, W. E. 1972. *Grillparzer: A Critical Introduction*. Cambridge: Cambridge UP.

Zimmerman, Michael. 1992. "Ontological Aestheticism: Heidegger, Jünger, and National Socialism." In *The Heidegger Case: On Philosophy and Politics*, ed. Tom Rockmore and Joseph Margolis. Philadelphia: Temple UP.

i ek, Slavoj. 1997. *The Plague of Fantasies*. London: Verso.

Index